SpringerBriefs in Electrical and Computer Engineering

Control, Automation and Robotics

Series editors

Tamer Başar
Antonio Bicchi
Miroslav Krstic

More information about this series at http://www.springer.com/series/10198

Alex S. Leong · Daniel E. Quevedo
Subhrakanti Dey

Optimal Control of Energy Resources for State Estimation Over Wireless Channels

 Springer

Alex S. Leong
Faculty of Electrical Engineering
  and Information Technology (EIM-E)
Paderborn University
Paderborn, Nordrhein-Westfalen
Germany

Subhrakanti Dey
Department of Engineering Science
Uppsala University
Uppsala, Uppsala Län
Sweden

Daniel E. Quevedo
Faculty of Electrical Engineering
  and Information Technology (EIM-E)
Paderborn University
Paderborn, Nordrhein-Westfalen
Germany

Springer Brief so author copyright

ISSN 2191-8112   ISSN 2191-8120   (electronic)
SpringerBriefs in Electrical and Computer Engineering
ISSN 2192-6786   ISSN 2192-6794   (electronic)
SpringerBriefs in Control, Automation and Robotics
ISBN 978-3-319-65613-7   ISBN 978-3-319-65614-4   (eBook)
DOI 10.1007/978-3-319-65614-4

Library of Congress Control Number: 2017949162

Mathematics Subject Classification (2010): 93E11, 93E20, 94A05, 90C39, 90C40

MATLAB® is a registered trademark of The MathWorks, Inc., 3 Apple Hill Drive, Natick, MA 01760-2098,
USA, http://www.mathworks.com.

Printed on acid-free paper

This Springer imprint is published by Springer Nature
The registered company is Springer International Publishing AG
The registered company address is: Gewerbestrasse 11, 6330 Cham, Switzerland

Contents

Chapter 1
Introduction

Recent years have seen increasing use of wireless technologies in diverse applications such as in environment monitoring, health care, and industrial monitoring and control. The convenience and cost savings due to the elimination of the need for wiring are readily apparent, for instance, the ease nowadays of setting up wireless networks in the home. Due to advances in micro-electro-mechanical technology, small and low-cost sensors with sensing, computation and wireless communication capabilities have become widely available. Such wireless sensors and actuators can be placed where wires cannot go, or where power sockets are not available. With this technology, ecosystems such as the Great Barrier Reef in Australia can be continuously monitored for pollution and to study the effects of climate change [1]. In health care, different body parts of a patient can be remotely monitored and doctors alerted if more medication or treatment is required [2].

New technologies such as cyber-physical systems [3, 4], smart cities [5, 6] and the Internet of Things [7, 8] have been envisioned, where many everyday objects are connected together (and also to the Internet) to form a network, potentially bringing further improvements to the quality of life. For example, using the location information of vehicles in the city can allow for improved traffic flow, alert drivers to areas of congestion and optimize the traffic light patterns for public transport. Autonomous vehicles, which will soon be available to the general public, can potentially lead to fewer accidents and increased traffic capacity. Buildings can be made more energy efficient by using more sophisticated control of the heating and cooling, ventilation and lighting, e.g. by using information on people's behaviour. Infrastructure such as water distribution networks can be continuously monitored, such that faults are detected and attended to quickly, leading to less wastage of resources. Air quality can be monitored with high resolution, alerting citizens and authorities when needed. In all these cases, communication with other sensors/devices or central authorities will be mainly over wireless channels.

© The Author(s) 2018   1
A.S. Leong et al., *Optimal Control of Energy Resources for State Estimation
Over Wireless Channels*, SpringerBriefs in Control, Automation and Robotics,
DOI 10.1007/978-3-319-65614-4_1

Fig. 1.1 Networked control system

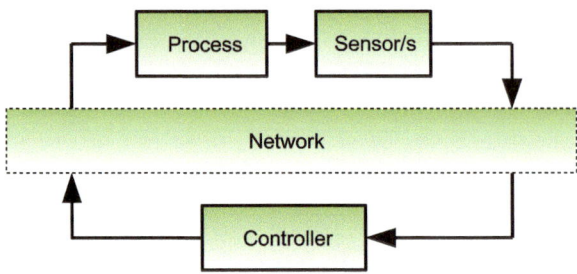

Networked Control Systems

In industrial monitoring and control applications, the driving force behind using wireless technology is its lower deployment and reconfiguration cost. Here, reliable operation is of vital importance, as a delay in implementing appropriate feedback control actions can drive a dynamical system to instability and cause substantial damage (both physical and financial). Nonetheless, even in such mission critical domains, the use of wireless technologies has become more prevalent, with industrial wireless sensor networks standards such as WirelessHART [9] being developed.

This has motivated significant research into networked control systems [10–13], where measurement and control signals are transmitted over channels (e.g. fading channels) or networks, see Fig. 1.1. In addition to bit rate limitations, these channels/networks introduce effects such as packet drops and delays. Research in networked control systems has included the derivation of conditions for stability and stabilizability, and methods for designing estimators and controllers which have a degree of robustness, in the presence of such effects. To derive these results, often the wireless environment has been abstracted into stochastic models such as i.i.d. (or Markovian) packet dropping links or regarded fading as multiplicative noise, which are then studied using mainly control-theoretic techniques. However, our view is that techniques and ideas from wireless communications itself, such as how to optimally manage energy resources in the presence of fading, can also be effectively utilized in the study and design of networked control systems. This book aims to demonstrate that making use of these additional techniques can provide significant performance gains.

Wireless Communications

Wireless channels, also known as fading channels, are inherently randomly time-varying, due to the small-scale effect of multipath, and larger scale effects such as path loss and shadowing by obstacles [14]. This can cause the transmitted signals to be attenuated, distorted, delayed or lost in ways which are difficult to predict, see Fig. 1.2. Signals transmitted by different sensors/devices over the wireless medium can also interfere with each other. Maintaining acceptable quality of service in such conditions is a challenging problem. Nevertheless, the goal of communicating with high reliability at ever higher data rates has been pursued extensively by the wireless

Fig. 1.2 Channel measurements taken at a paper mill [15]

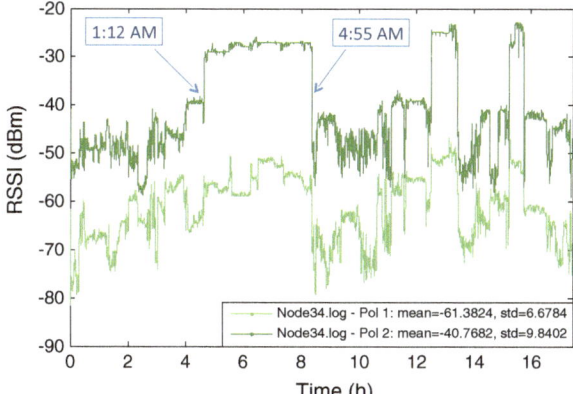

communications community over the last 30 years, and many novel ideas have been developed.

In particular, power control is a key enabling technology for wireless communications. It serves to compensate for the time-varying channel gains, to provide quality of service guarantees to users and also to increase spectral efficiency [14]. Energy harvesting technologies [16, 17], where the sensor can recharge their batteries by extracting energy from the surrounding environment, can dramatically increase the lifetime of nodes in wireless sensor networks, and will be fundamental in the implementation of self-sustaining cyber-physical systems and the Internet of Things. Coding schemes can improve the reliability of the transmitted information, with network coding even able to increase throughput [18, 19]. Multi-hop networks and cooperative communications via the use of relays [20] have been identified as some of the key enabling technologies for fifth generation (5G) mobile networks [21]. These ideas, just to name a few, have not received much attention in the study of networked control systems so far, while we believe that they can, and should, be fruitfully utilized.

Contributions and Scope of This Book

One of the goals of this book is to bring closer together the wireless communications and control literature, by introducing wireless communications techniques and ideas into the study and design of networked control systems. For that purpose, the focus is on state estimation problems where sensor measurements (or related quantities such as local state estimates or innovations) are transmitted over wireless links to a central observer, see Fig. 1.3. State estimation of dynamical systems is important in areas such as environment monitoring and tracking. Furthermore, estimated state feedback control forms a central part of contemporary control systems [22]. Indeed, some of the topics studied in this book have also been extended to feedback control [23–26].

Fig. 1.3 Networked
estimation

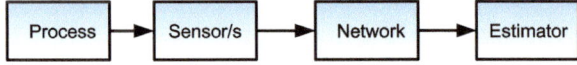

The approach taken in this book is to utilize some of the techniques and ideas that have been developed in wireless communications for energy[1] resource management, in order to improve the performance of the estimator when transmission occurs over wireless packet dropping links. Many previous works have studied Kalman filtering and control over packet dropping links, but where the energy usage is not explicitly taken into account [27, 28]. Some works in the networked estimation and control literature have considered fading [29–33], but apart from [32] the fading is often treated as unknown multiplicative noise. This can lead to conservative designs, since in practice, often knowledge of the fading channel gains is available[2] at the receiver (or transmitter). Thus, in this book, we will assume that the fading channel gains are known and can be exploited, e.g. by using power control [14] to enhance system performance.

Energy harvesting based rechargeable batteries or storage devices can offer significant advantages in the deployment of large-scale wireless sensor and actuator networks. These devices, when integrated into the sensor/actuator nodes, provide freedom from the task of periodically having to replace batteries, and open the possibility for sensors to operate in a self-sustaining manner. Recent research on energy harvesting has largely focused on resource allocation for wireless communication systems design, optimizing communication objectives such as maximizing throughput or minimizing transmission delay [35–37]. In contrast, in this book, we directly optimize estimation objectives such as minimizing the expected estimation error covariance.

This book focuses on performance optimization of networked estimation systems, which goes beyond the notion of stability/stabilizability considered in many papers. As mentioned previously, two fundamental aspects in wireless communications are fading and interference [14]. In this book, we directly deal with fading, while interference is indirectly dealt with in our formulation via the particular abstraction of packet loss that can consider interference by using higher packet loss probabilities. In terms of robustness issues, our work in power control addresses robustness with respect to the time-varying behaviour of the fading channels. However, robustness in terms of model imperfections lies beyond the scope of this book and will not be considered.

Book Outline

Chapter 2 deals with power allocation for state estimation of discrete-time linear dynamical systems. The sensors transmit measurements over a packet dropping channel, where the probability of successful packet reception is time-varying and

---

[1] We measure energy on a per channel use basis and will refer to energy and power interchangeably.

[2] In wireless communication, this is referred to as having channel state information at the receiver or transmitter [34].

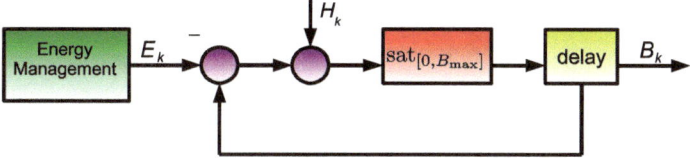

Fig. 1.4 Energy harvesting

Fig. 1.5 Event triggered
estimation

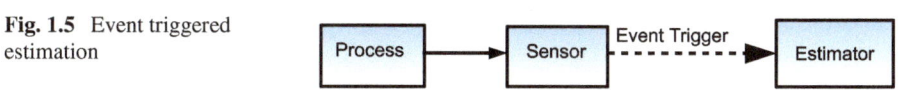

depends on the instantaneous fading channel gain and transmit power used. We first consider a transmission power control problem to minimize a linear combination of the expected error covariance and expected energy usage. We then study an optimal transmission power control problem where energy is randomly harvested from the environment. Here, the transmission energy is constrained by the battery level (which depends on the energy used in previous time steps and the energy harvested in the meantime), see Fig. 1.4.

Chapter 3 considers event-triggered estimation, see Fig. 1.5, where a sensor will transmit local state estimates to a remote estimator only when certain events occur, e.g. if the estimation quality has deteriorated sufficiently. This can save energy while still maintaining a certain level of performance. In particular, we study the case where the transmission decisions are obtained from the solution to an optimization problem that minimizes a linear combination of the expected error covariance and expected energy usage. We derive structural results which show that the optimal policy is of threshold type, i.e. transmit if and only if the error covariance at the remote estimator exceeds a certain threshold. This provides a rigorous justification of variance-based threshold policies proposed in [38]. We then consider the problem of minimizing the expected error covariance subject to energy harvesting constraints, where a transmission can occur only if there is sufficient energy in the battery. We show that the optimal policies have a threshold structure in both the error covariance and the battery level.

Chapter 4 studies a design problem which arises in the context of optimal transmission strategies for remote state estimation. We consider the case where a sensor can either transmit its local state estimate or its local innovations. While transmitting local estimates will give improved performance at the remote estimator, often it will also have a larger variance and require more energy to transmit. This raises the issue of finding a transmission strategy that optimizes a linear combination of the expected error covariance and expected energy usage. For scalar systems, it turns out that (similar to Chap. 3) the optimal strategy also has a threshold structure, where one transmits the estimates if the error covariance exceeds a certain threshold, and transmits the innovations otherwise.

Fig. 1.6 Multi-hop network

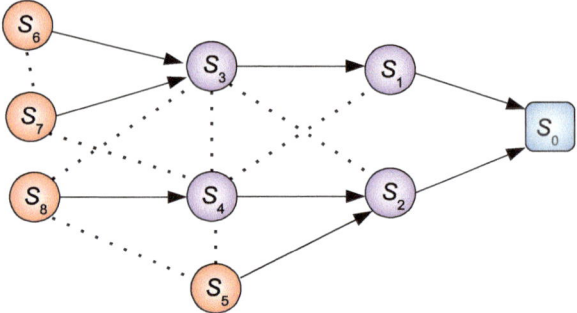

Chapter 5 focuses on remote state estimation problems over multi-hop networks, see Fig. 1.6. Here, we show that performance benefits can be obtained if one adopts more advanced communication techniques such as network coding [18, 19], relays [20], and rerouting [39]. We first consider a set-up where sensors can transmit both directly to the remote estimator or via intermediate relays. We consider different operations that the relay can perform such as forwarding of transmissions or network coding operations, and optimize over the relay operations and transmission powers. Next, we consider the problem of reconfiguring the topology of (or *rerouting*) a multi-hop network, in order to respond to time variations in the wireless channel conditions. Optimal and suboptimal methods for reconfiguring the network are proposed, and their performance compared.

References

1. GBROOS-data, http://data.aims.gov.au/gbroos
2. M. Chen, S. Gonzalez, A. Vasilakos, H. Cao, V.C.M. Leung, Body area networks: a survey. Mobile Netw. Appl. **16**(2), 171–193 (2011)
3. R. Rajkumar, I. Lee, L. Sha, J. Stankovic, Cyber-physical systems: the next computing revolution, in *Proceedings of the ACM Design Automation Conference*, Anaheim, CA (2010), pp. 731–736
4. R. Poovendran, K. Sampigethaya, S.K.S. Gupta, I. Lee, K.V. Prasad, D. Corman, J.L. Paunicka (eds.), Special issue on cyber-physical systems. Proc. IEEE **100**(1) (2012)
5. H. Chourabi, T. Nam, S. Walker, J.R. Gil-Garcia, S. Mellouli, K. Nahon, T.A. Pardo, H.J. Scholl, Understanding smart cities: an integrative framework, in *Proceedings of the HICSS*, Maui, HI (2012), pp. 2289–2297
6. IEEE Smart Cities Initiative, http://smartcities.ieee.org
7. L. Atzori, A. Iera, G. Morabito, The internet of things: a survey. Comput. Netw. **54**(15), 2787–2805 (2010)
8. IEEE Internet of Things Initiative, http://iot.ieee.org
9. HART Communication Foundation, http://en.hartcomm.org/
10. P. Antsaklis, J. Baillieul (eds.), Special issue on technology of networked control systems. Proc. IEEE **95**(1) (2007)
11. M. Franceschetti, T. Javidi, P.R. Kumar, S. Mitter, D. Teneketzis (eds.), Special issue on control and communications. IEEE J. Sel. Areas Commun. **26**(4) (2008)

12. J. Chen, K.H. Johansson, S. Olariu, I.C. Paschalidis, I. Stojmenovic (eds.), Special issue on wireless sensor and actuator networks. IEEE Trans. Autom. Control **56**(10) (2011)
13. K.H. Johansson, G.J. Pappas, P. Tabuada, C.J. Tomlin (eds.), Special issue on control of cyber-physical systems. IEEE Trans. Autom. Control **59**(12) (2014)
14. D.N.C. Tse, P. Viswanath, *Fundamentals of Wireless Communication* (Cambridge University Press, Cambridge, 2005)
15. P. Agrawal, A. Ahlén, T. Olofsson, M. Gidlund, Long term channel characterization for energy efficient transmission in industrial environments. IEEE Trans. Commun. **62**(8), 3004–3014 (2014)
16. M.M. Tentzeris, A. Georgiadis, L. Roselli (eds.), Special issue on energy harvesting and scavenging. Proc. IEEE **102**(11) (2014)
17. S. Ulukus, E. Erkip, P. Grover, K. Huang, O. Simeone, A. Yener, M. Zorzi (eds.), Special issue on wireless communications powered by energy harvesting and wireless energy transfer. IEEE J. Sel. Areas Commun. **33**(3) (2015)
18. R. Ahlswede, N. Cai, S.-Y.R. Li, R.W. Yeung, Network information flow. IEEE Trans. Inf. Theory **46**(7), 1204–1216 (2000)
19. S. Katti, H. Rahul, W. Hu, D. Katabi, M. Médard, J. Crowcroft, XORs in the air: practical wireless network coding. IEEE/ACM Trans. Netw. **16**(3), 497–510 (2008)
20. Y. Hua, D.W. Bliss, S. Gazor, Y. Rong, Y. Sung (eds.), Special issue on theories and methods for advanced wireless relays. IEEE J. Sel. Areas Commun. **30**(8) (2012)
21. T.S. Rappaport, R.W. Heath Jr., R.C. Daniels, J.N. Murdock, *Millimeter Wave Wireless Communications* (Prentice Hall, Upper Saddle River, 2014)
22. K.J. Aström, R.M. Murray, *Feedback Systems: An Introduction for Scientists and Engineers* (Princeton University Press, Princeton, 2008)
23. S. Knorn, S. Dey, Optimal energy allocation for linear control with packet loss under energy harvesting constraints. Automatica **77**, 259–267 (2017)
24. S. Dey, A. Chiuso, L. Schenato, Feedback control over lossy SNR-limited channels: linear encoder-decoder-controller design. IEEE Trans. Autom. Control **62**(6), 3054–3061 (2017)
25. A.S. Leong, D.E. Quevedo, T. Tanaka, S. Dey, A. Ahlén, Event-based transmission scheduling and LQG control over a packet dropping link, in *Proceedings of the IFAC World Congress*, Toulouse, France (2017), pp. 9275–9280
26. A.S. Leong, S. Dey, D.E. Quevedo, Transmission scheduling for remote state estimation and control with an energy harvesting sensor (2017), submitted for publication
27. B. Sinopoli, L. Schenato, M. Franceschetti, K. Poolla, M.I. Jordan, S.S. Sastry, Kalman filtering with intermittent observations. IEEE Trans. Autom. Control **49**(9), 1453–1464 (2004)
28. L. Schenato, B. Sinopoli, M. Franceschetti, K. Poolla, S.S. Sastry, Foundations of control and estimation over lossy networks. Proc. IEEE **95**(1), 163–187 (2007)
29. N. Elia, Remote stabilization over fading channels. Syst. Control Lett. **54**, 237–249 (2005)
30. Y. Mostofi, R.M. Murray, To drop or not to drop: design principles for Kalman filtering over wireless fading channels. IEEE Trans. Autom. Control **54**(2), 376–381 (2009)
31. N. Xiao, L. Xie, L. Qiu, Feedback stabilization of discrete-time networked systems over fading channels. IEEE Trans. Autom. Control **57**(9), 2176–2189 (2012)
32. K. Gatsis, A. Ribeiro, G.J. Pappas, Optimal power management in wireless control systems. IEEE Trans. Autom. Control **59**(6), 1495–1510 (2014)
33. H. Dong, Z. Wang, S.X. Ding, H. Gao, On $\mathcal{H}_\infty$ estimation of randomly occurring faults for a class of nonlinear time-varying systems with fading channels. IEEE Trans. Autom. Control **61**(2), 479–484 (2016)
34. A.F. Molisch, *Wireless Communications*, 2nd edn. (Wiley, New York, 2011)
35. V. Sharma, U. Mukherji, V. Joseph, S. Gupta, Optimal energy management policies for energy harvesting sensor nodes. IEEE Trans. Wirel. Commun. **9**(4), 1326–1336 (2010)
36. O. Ozel, K. Tutuncuoglu, J. Yang, S. Ulukus, A. Yener, Transmission with energy harvesting nodes in fading wireless channels: optimal policies. IEEE J. Sel. Areas Commun. **29**(8), 1732–1743 (2011)

37. C.K. Ho, R. Zhang, Optimal energy allocation for wireless communications with energy harvesting constraints. IEEE Trans. Signal Process. **60**(9), 4808–4818 (2012)
38. S. Trimpe, R. D'Andrea, Event-based state estimation with variance-based triggering. IEEE Trans. Autom. Control **59**(12), 3266–3281 (2014)
39. J.F. Kurose, K. Ross, *Computer Networking: A Top-Down Approach*, 6th edn. (Pearson, Boston, 2012)

Chapter 2
Optimal Power Allocation for Kalman Filtering over Fading Channels

Kalman filtering with random packet drops has been studied extensively since the work of [1], which showed that for i.i.d. Bernoulli packet drops, there exists a critical threshold such that if the packet arrival rate exceeds this threshold, then the expected error covariance remains bounded, but diverges otherwise. This work has been extended in various directions such as: multiple sensors [2, 3], further characterizations of the critical threshold [4, 5], probabilistic notions of performance [6, 7], performing local processing before transmission [8], consideration of delays [9] and Markovian packet drops [10, 11].

As mentioned in Chap. 1, in wireless communications, power control is regularly used to improve system performance and reliability [12, 13]. The primary focus of the previously mentioned works is on deriving conditions for stability of the estimator, and power control is not explicitly considered. However, power control can also be used in Kalman filtering to improve the estimator stability and estimation performance. For Kalman filtering over continuous fading channels, the use of power control for outage minimization and expected error covariance minimization has been studied in [14]. The works of [15, 16] consider the use of power control at the sensor over a continuous fading channel, with the data being sent over this channel after digital modulation, which would then give a corresponding packet loss probability dependent on the transmit power at the sensor. Power allocation using model predictive control techniques is considered in [15], while optimal power allocation schemes to guarantee stability are investigated in [16].

In conventional wireless communication systems, the sensors have access either to a fixed energy supply or have batteries that may be easily recharged/replaced. In contrast, when energy harvesting capabilities are available, then the sensors can recharge their batteries by collecting energy from the environment, e.g. solar, thermal, mechanical vibrations, or electromagnetic radiation [17, 18]. In the context of wireless sensor networks, the use of energy harvesting may be especially useful, e.g. in remote locations with restricted access to an energy supply, and even mandatory

© The Author(s) 2018
A.S. Leong et al., *Optimal Control of Energy Resources for State Estimation Over Wireless Channels*, SpringerBriefs in Control, Automation and Robotics, DOI 10.1007/978-3-319-65614-4_2

where it is dangerous or impossible to change the batteries. The amount of energy harvested is random as most renewable energy sources are unreliable. Clearly, the energy expenditure at every time slot is constrained by the amount of stored energy currently available. This, however, complicates the design of suitable transmission power allocation policies. Communication schemes for optimizing throughput or minimizing transmission delay for transmitters with energy harvesting capability have been studied in [19–23], while a remote estimation problem with an energy harvesting sensor was considered in [24], which minimized a cost consisting of both the distortion and the number of sensor transmissions.

In this chapter, we adopt the channel model of [15, 16], but instead of using power allocation to achieve filter stability, we are interested in the use of power allocation to improve the estimation performance of the Kalman filter. Section 2.1 first studies optimal power allocation for sensors without energy harvesting capabilities. Here, we focus on minimizing the trace of the expected error covariance subject to an average transmit power constraint. The problem is formulated as a Markov decision process (MDP) problem that can be solved numerically with dynamic programming techniques. Two simpler suboptimal schemes are also investigated, namely a constant power allocation scheme and a truncated channel inversion policy. Section 2.2 then investigates the situation with an energy harvesting sensor. An important issue is to address the trade-off between the use of available stored energy to improve the current transmission reliability (and thus state estimation accuracy), or the storing of energy for future transmissions which may be affected by higher packet loss probabilities due to severe fading. The optimal transmission energy allocation policies are obtained by the use of dynamic programming techniques. Using the concept of submodularity [25], the structure of the optimal transmission energy policies is also studied.

2.1 Optimal Power Allocation for Remote State Estimation

2.1.1 System Model

A diagram of the system model for this section is given in Fig. 2.1. Consider a linear system

$$x_{k+1} = Ax_k + w_k \tag{2.1}$$

where $x_k \in \mathbb{R}^n$, and w_k is i.i.d. Gaussian with zero mean and covariance matrix $Q > 0$.[1] The sensor makes a measurement

$$y_k = Cx_k + v_k \tag{2.2}$$

---

[1] We say that a matrix $X > 0$ if X is positive definite, and $X \geq 0$ if X is positive semi-definite.

Fig. 2.1 Transmission
power control for remote
state estimation

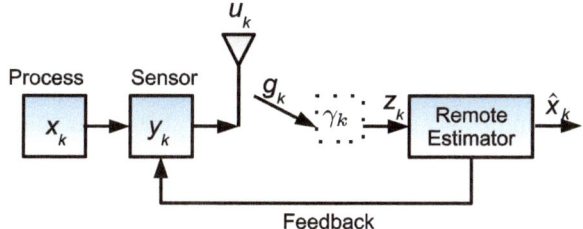

where $y_k \in \mathbb{R}^m$, and v_k is i.i.d. Gaussian with zero mean and covariance matrix $R > 0$. We assume that the pair (A, C) is detectable and the pair $(A, Q^{1/2})$ is stabilizable.

The measurement is then sent to a remote estimator over a packet dropping link, which can be modelled as

$$z_k = \gamma_k y_k,$$

where z_k is the quantity received at the remote estimator. Here, the measurement y_k is assumed to be encoded to form a single packet, and $\gamma_k = 1$ denotes that the measurement packet is received (i.e. correctly decoded), while $\gamma_k = 0$ denotes that the packet is lost (i.e. corrupted).[2]

Kalman Filter with Random Packet Drops

In order to estimate the state x_k, the remote estimator runs a Kalman filter, which also takes into account the random packet drops [1]. The Kalman filter state estimates and error covariances are defined as:

$$\hat{x}_{k|k} = \mathbb{E}[x_k | z_0, \ldots, z_k, \gamma_0, \ldots, \gamma_k]$$
$$\hat{x}_{k+1|k} = \mathbb{E}[x_{k+1} | z_0, \ldots, z_k, \gamma_0, \ldots, \gamma_k]$$
$$P_{k|k} = \mathbb{E}[(x_k - \hat{x}_{k|k})(x_k - \hat{x}_{k|k})^T | z_0, \ldots, z_k, \gamma_0, \ldots, \gamma_k]$$
$$P_{k+1|k} = \mathbb{E}[(x_{k+1} - \hat{x}_{k+1|k})(x_{k+1} - \hat{x}_{k+1|k})^T | z_0, \ldots, z_k, \gamma_0, \ldots, \gamma_k].$$

The Kalman filtering equations with packet drops are given by:

$$\begin{aligned}
\hat{x}_{k|k} &= \hat{x}_{k|k-1} + \gamma_k K_k (y_k - C\hat{x}_{k|k-1}) \\
\hat{x}_{k+1|k} &= A\hat{x}_{k|k} \\
P_{k|k} &= P_{k|k-1} - \gamma_k P_{k|k-1} C^T (C P_{k|k-1} C^T + R)^{-1} C P_{k|k-1} \\
P_{k+1|k} &= A P_{k|k} A^T + Q,
\end{aligned} \tag{2.3}$$

where $K_k = P_{k|k-1} C^T (C P_{k|k-1} C^T + R)^{-1}$. In this chapter, we will also use the shorthand $P_k \triangleq P_{k|k-1}$. Then $\{P_k\}$ satisfies

$$P_{k+1} = A P_k A^T + Q - \gamma_k A P_k C^T (C P_k C^T + R)^{-1} C P_k A^T.$$

---

[2]In practice this can be determined using simple error detecting codes.

Packet Drop Model

In this chapter, we will adopt a model from [15, 16] for the packet loss process $\{\gamma_k\}$ that is governed by a time-varying wireless fading channel $\{g_k\}$ and the sensor transmit power control $\{u_k\}$ over this channel. In this model, the conditional packet reception probabilities are given by

$$\mathbb{P}(\gamma_k = 1 | g_k, u_k) \triangleq f(g_k u_k) \qquad (2.4)$$

where $f(.) : [0, \infty) \to [0, 1]$ is a monotonically increasing continuous function. The form of $f(.)$ will depend on the particular digital modulation scheme being used [26], see e.g. (2.12) for the case of binary phase shift keying (BPSK) transmission.

We will consider the case where $\{g_k\}$ is an i.i.d. block fading process [27], where the channel remains constant over a fading block (representing the coherence time of the channel [28]) but can vary from block to block in an i.i.d. manner.

Kalman Filter Stability

We assume that channel state information (CSI) is available at the remote estimator, such that the remote estimator knows the values of the channel gains g_k at time k.[3] Since CSI is assumed to be available, we will allow the sensor transmit power u_k to depend on both g_k and P_k. Note that if the energy allocation u_k is computed based on the estimation error covariance (and not the state x_k), then the optimal estimator is still given by the Kalman filter (2.3). In the next section, we consider optimal power allocation to minimize the trace of the expected error covariance. Due to limited computational resources at the sensor, the optimal sensor transmit powers are computed at the remote estimator and fed back to the sensor.[4]

Using techniques from [29], we can obtain the following sufficient condition for stability of the Kalman filter, for power control schemes $\{u_k\}$ which are allowed to depend on the channel gains g_k and error covariances P_k.

Theorem 2.1 *Let $\|A\|$ denote the spectral norm of A. If there exists an $r \in [0, 1)$ such that:*

$$\mathbb{P}(\gamma_k = 1) \geq 1 - \frac{r}{\|A\|^2}, \quad \forall k \in \mathbb{N},$$

then $\{P_k\}$ satisfies

$$\mathbb{E}[\mathrm{tr}(P_k)] \leq \alpha r^k + \beta, \quad \forall k \in \mathbb{N} \qquad (2.5)$$

for some $\alpha, \beta \in \mathbb{R}$.

---

[3]In practice, this can be achieved by periodically sending pilot signals either from the sensor to the remote estimator to allow the remote estimator to estimate the channel, or from the remote estimator to the sensor under channel reciprocity.

[4]In wireless communications, online computation of powers at the base station, which is then fed back to the mobile transmitters, is commonly done in practice [12], at time scales on the order of milliseconds.

2.1.2 Optimal Power Allocation

The problem we consider in this subsection is to determine the optimal sensor transmit power allocation, in order to minimize the trace of the expected error covariance subject to an average transmit power constraint $\mathcal{P}$, i.e. we are interested in solving

$$
\begin{aligned}
&\min_{\{u_k\}} \limsup_{K\to\infty} \frac{1}{K} \sum_{k=0}^{K-1} \mathbb{E}[\mathrm{tr}(P_{k+1})] \\
&\text{s.t. } \limsup_{K\to\infty} \frac{1}{K} \sum_{k=0}^{K-1} \mathbb{E}[u_k] \leq \mathcal{P}.
\end{aligned}
\tag{2.6}
$$

Remark 2.1 When the system matrix A is unstable (i.e. has eigenvalues outside the unit circle), Kalman filtering with packet losses can have unbounded expected error covariances in certain situations [1]. This then raises the question as to whether problem (2.6) is well posed. In [16], we studied the problem of determining the minimum average power required for guaranteeing that Theorem 2.1 is satisfied. Choosing $\mathcal{P}$ in the average power constraint of problem (2.6) to be greater than this minimum average power (see [16] for details on how to compute this minimum average power) will be sufficient to make (2.6) well posed.

The optimization problem (2.6) can be regarded as a constrained average cost Markov decision process (MDP) problem [30] with (P_k, g_k) as the 'state' and u_k as the 'action' of the MDP. To solve this problem, we will use a Lagrangian technique similar to [14, 30, 31] that considers instead the following unconstrained MDP problem:

$$
\begin{aligned}
&\min_{\{u_k\}} \limsup_{K\to\infty} \frac{1}{K} \sum_{k=0}^{K-1} \mathbb{E}[\mathrm{tr}(P_{k+1}) + \beta u_k] \\
&= \min_{\{u_k\}} \limsup_{K\to\infty} \frac{1}{K} \sum_{k=0}^{K-1} \mathbb{E}[\mathbb{E}[\mathrm{tr}(P_{k+1})|P_k, g_k, u_k] + \beta u_k],
\end{aligned}
\tag{2.7}
$$

where $\beta \geq 0$ specifies the trade-off between the average transmit power and expected error covariance. Solving (2.7) for different values of β will then correspond to minimizing the trace of the expected error covariance for different average transmit power constraints in (2.6).

The average cost optimality equation or Bellman equation [32] associated with problem (2.7) can be written as

$$
\begin{aligned}
\rho + h(P_k, g_k) = \min_{u_k} \Big[&\mathbb{E}[\mathrm{tr}(P_{k+1})|P_k, g_k, u_k] + \beta u_k \\
&+ \int_{g_{k+1}, P_{k+1}} h(P_{k+1}, g_{k+1}) F(d(P_{k+1}, g_{k+1})|P_k, g_k, u_k) \Big],
\end{aligned}
\tag{2.8}
$$

where ρ is the optimal average cost per stage, h the differential cost and F the probability transition law of (P_k, g_k).

We first show that there exist stationary solutions to the MDP (2.7). We will make the following additional assumption:

Assumption 2.1.1 The range of u_k is bounded, i.e. $u_k \in [0, u_{max}]$, $\forall k$.

Such an assumption is obviously justified from a practical point of view.

Lemma 2.1 *Under Assumption 2.1.1, there exists a stationary solution to the Bellman equation (2.8) which solves the MDP (2.7).*

Proof The proof involves verifying the conditions from [33] that guarantee the existence of stationary solutions for MDPs with Borel state and action spaces. The verification of these conditions is very similar to the proof of Lemma 3 in [14], see also the proof of Theorem 2.3 in the appendix to this chapter. The details are omitted for brevity. □

For computational purposes, the Bellman equation can be further simplified as follows:

$$
\rho + h(P_k, g_k)
$$
$$
= \min_{u_k} \left[\mathbb{E}[\text{tr}(P_{k+1})|P_k, g_k, u_k] + \beta u_k \int_{g_{k+1}, P_{k+1}} h(P_{k+1}, g_{k+1}) F(d(P_{k+1}, g_{k+1})|P_k, g_k, u_k) \right]
$$
$$
= \min_{u_k} \left\{ \text{tr}(A P_k A^T + Q) + \beta u_k - f(g_k u_k) \text{tr}\left(A P_k C^T (C P_k C^T + R)^{-1} C P_k A^T \right) \right.
$$
$$
\left. + \int_{P_{k+1}, g_{k+1}} h(P_{k+1}, g_{k+1}) F(d(P_{k+1}, g_{k+1})|P_k, g_k, u_k) \right\}
$$
$$
\overset{(a)}{=} \min_{u_k} \left\{ \text{tr}(A P_k A^T + Q) + \beta u_k - f(g_k u_k) \text{tr}\left(A P_k C^T (C P_k C^T + R)^{-1} C P_k A^T \right) \right.
$$
$$
\left. + \int_{P_{k+1}, g_{k+1}} h(P_{k+1}, g_{k+1}) F(d P_{k+1}|P_k, g_k, u_k) F(d g_{k+1}) \right\}
$$
$$
\overset{(b)}{=} \min_{u_k} \left\{ \text{tr}(A P_k A^T + Q) + \beta u_k - f(g_k u_k) \text{tr}\left(A P_k C^T (C P_k C^T + R)^{-1} C P_k A^T \right) \right.
$$
$$
+ \int_{g_{k+1}} \left[h(A P_k A^T + Q, g_{k+1})(1 - f(g_k u_k)) \right.
$$
$$
\left. + h\left(A P_k A^T + Q - A P_k C^T (C P_k C^T + R)^{-1} C P_k A^T, g_{k+1} \right) f(g_k u_k) \right] F(d g_{k+1}) \right\}
$$
$$
\tag{2.9}
$$

where (a) follows from the fact that g_{k+1} is independent of P_{k+1}, and (b) follows from writing out the conditional expectation $\mathbb{E}[h(P_{k+1}, g_{k+1})|P_k, g_k, u_k]$. For numerical implementation, a discretized version of the Bellman equation (2.9) can then be solved using, e.g. the relative value iteration algorithm [32] to find solutions to the MDP (2.7).

Remark 2.2 The discretized solution is, strictly speaking, a suboptimal approximation to the true optimal solution, however, the use of discretization is generally unavoidable for MDPs with continuous state and action spaces. As the number of discretization levels increases, the discretized solution usually converges to the optimal solution [34].

Now let $\mathfrak{p}^*(u)$ be the minimum trace of the expected error covariance such that the average transmit power is less than u. By solving the MDP (2.7) for different values of β, one can obtain points of the function $\mathfrak{p}^*(u)$, corresponding to different trade-offs between the average transmit power and trace of the expected error covariance, see Fig. 2.2. We have the following characterization of the function $\mathfrak{p}^*(u)$:

Lemma 2.2 *Suppose $f(.)$ in (2.4) is a strictly concave function. Then $\mathfrak{p}^*(u)$ is a decreasing strictly convex function of u.*

Proof See Appendix.

An example of a strictly concave $f(.)$ is given by (2.12) in Sect. 2.1.4. Using Lemma 2.2, one can conclude from the theory of Pareto optimality that all points on the curve $\mathfrak{p}^*(u)$ can be obtained by solving the MDP (2.7) for an appropriate choice of β, see [35, 36] for further details.

2.1.3 Suboptimal Power Allocation Policies

The optimal solution considered in the previous section requires the solution of an MDP, which is computationally demanding, particularly for vector systems. In this section, we consider two suboptimal policies which are simpler to compute and implement than the optimal solution of Sect. 2.1.2.

Constant Power Allocation

One very simple scheme is to use constant power allocation, where $u_k = u_{const}$, $\forall k$. With this policy, the conditional packet reception probabilities $f(g_k u_{const})$ will depend only on the channel gain g_k.

Truncated Channel Inversion

Another suboptimal scheme is based on the concept of channel inversion, which is a simple but quite commonly used technique in wireless communications, that attempts to invert the channel at every time instance to maintain a constant quality of service. However, it is known that for certain fading distributions such as Rayleigh fading, channel inversion actually requires infinite average power, so some modifications to the scheme such as truncation (where channel inversion is only carried out if the channel gain is sufficiently large) are necessary [37]. The power allocation policy we consider here is of the following form:

$$u_k = \begin{cases} \alpha/g_k, & \text{if } g_k > g^* \\ \alpha/g^*, & \text{otherwise} \end{cases} \tag{2.10}$$

where α and g^* are values which can be designed. This scheme inverts the channel g_k and multiplies it by a gain α if g_k is greater than some threshold g^*, otherwise it transmits with the constant power $\frac{\alpha}{g^*}$. The average transmit power using this scheme is

$$\mathbb{E}[u_k] = \int_{g^*}^{\infty} \frac{\alpha}{g_k} F(dg_k) + \int_0^{g^*} \frac{\alpha}{g^*} F(dg_k)$$
$$= \alpha E(g^*) + \frac{\alpha}{g^*} F_G(g^*), \forall k$$

where

$$E(g^*) \triangleq \int_{g^*}^{\infty} \frac{1}{g_k} F(dg_k),$$

and $F_G(.)$ is the cumulative distribution function of g_k. For instance, if $g_k \sim \text{Exp}(1)$, which is an example of Rayleigh fading [28], we have $E(g^*) = \int_{g^*}^{\infty} \exp(-g_k)/g_k \, dg_k = E_1(g^*)$ (i.e. the exponential integral), and $F_G(g^*) = 1 - \exp(-g^*)$.

In terms of the packet loss process $\{\gamma_k\}$, under this power allocation scheme, $\gamma_k = 1$ with conditional probability $f(\alpha)$ when $g_k > g^*$, and $\gamma_k = 1$ with conditional probability $f(\frac{\alpha g_k}{g^*})$ when $g_k \leq g^*$. That is, we have

$$\gamma_k = \begin{cases} 1, & \text{w.p. } f(\alpha)(1 - F_G(g^*)) + \int_0^{g^*} f\left(\frac{\alpha g_k}{g^*}\right) F(dg_k) \\ 0, & \text{w.p. } (1 - f(\alpha))(1 - F_G(g^*)) + \int_0^{g^*} \left(1 - f\left(\frac{\alpha g_k}{g^*}\right)\right) F(dg_k). \end{cases}$$

Therefore, using this scheme, γ_k becomes an i.i.d. Bernoulli process with probability of successful packet reception $f(\alpha)(1 - F_G(g^*)) + \int_0^{g^*} f(\frac{\alpha g_k}{g^*}) F(dg_k)$.

As the values α and g^* can be chosen by us, we can optimize α and g^* to minimize the trace of the expected error covariance subject to an average power constraint, i.e. solving problem (2.6) but with u_k restricted to be of the form (2.10). For i.i.d. packet losses, it is known that the expected error covariance is a decreasing function of the packet reception probability [1]. Hence, the problem is equivalent to minimizing the probability of packet loss subject to an average power constraint $\mathscr{P}$, i.e.

$$\min_{\alpha, g^*} (1 - f(\alpha))(1 - F_G(g^*)) + \int_0^{g^*} \left(1 - f\left(\frac{\alpha g_k}{g^*}\right)\right) F(dg_k)$$
$$\text{s.t. } \alpha E(g^*) + \frac{\alpha}{g^*} F_G(g^*) = \mathscr{P}. \tag{2.11}$$

We can further simplify problem (2.11) by rearranging the constraint to express α in terms of g^*, i.e.

Fig. 2.2 Average transmit power versus expected error covariance

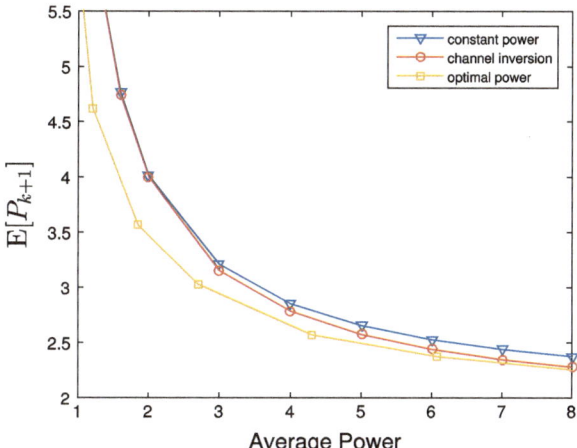

$$\alpha = \frac{\mathscr{P}}{E(g^*) + \frac{1}{g^*} F_G(g^*)}.$$

The optimization problem (2.11) then becomes a one-dimensional line search over g^*, which can be easily solved numerically.

2.1.4   Numerical Studies

We present here numerical results for a scalar system with parameters $A = 1.2$, $C = 1$, $Q = 1$, $R = 1$. We consider the case where the digital communication uses binary phase shift keying (BPSK) transmission [26] with b bits per packet, so that we have

$$\mathbb{P}(\gamma_k = 1 | g_k, u_k) = f(g_k u_k) = \left(\int_{-\infty}^{\sqrt{g_k u_k}} \frac{1}{\sqrt{2\pi}} e^{-t^2/2} dt \right)^b \qquad (2.12)$$

One can verify that $f(.)$ is a strictly concave function for $b \in \{1, 2, 3, 4, 5\}$. In the simulations below we use $b = 4$. The fading channel is taken to be Rayleigh [28], so that g_k is exponentially distributed with p.d.f.

$$p(g_k) = \frac{1}{\bar{g}} \exp(-g_k/\bar{g}), g_k \geq 0$$

with $\bar{g}$ being its mean. Here, we will use $\bar{g} = 1$. In solving the Bellman equation (2.9), we use 50 discretization points for each of the quantities P_k, g_k, u_k, see Remark 2.2.

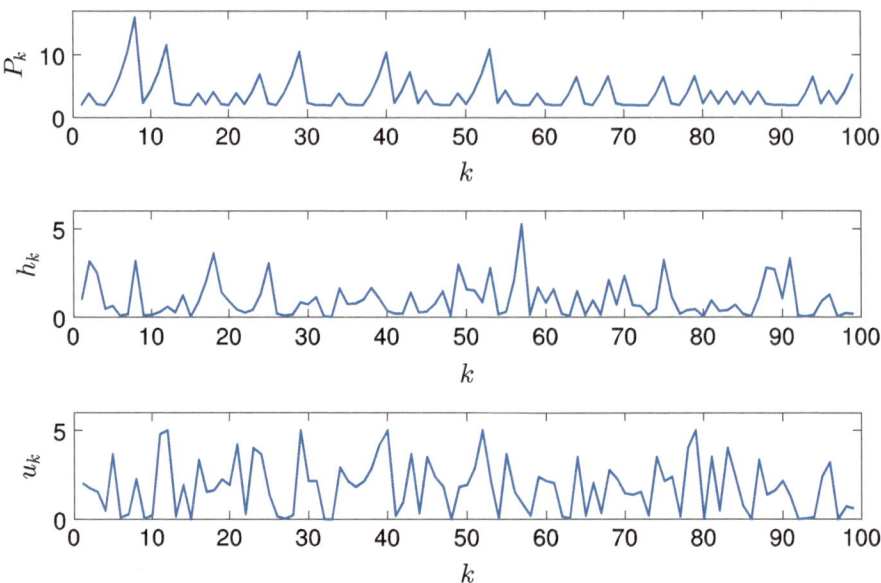

Fig. 2.3 Optimal power allocations

In Fig. 2.2, we plot the average transmit power versus expected error covariance trade-off, for the cases of optimal power allocation of Sect. 2.1.2, and the constant power allocation and truncated channel inversion policies of Sect. 2.1.3. We see that optimal power allocation has significant performance gains over the simpler subopti-mal policies of Sect. 2.1.3 for low average transmit powers, with the performance of the constant power allocation and channel inversion policies being almost identical. While for higher average transmit powers, the truncated channel inversion policy has performance approaching that of the optimal power allocation policy.

In Fig. 2.3, we plot a single simulation run of P_k and g_k, together with the cor-responding optimal power allocations u_k. We can see that in the optimal power allocation scheme, the allocated powers will depend on both the current channel gain g_k and error covariance P_k. The allocated power u_k tends to be higher when the error covariance P_k is larger, provided the corresponding channel gain g_k is not too small.

2.2 Optimal Power Allocation with Energy Harvesting

A diagram of the system architecture for this section is shown in Fig. 2.4. The model for the process (2.1) and (2.2) and packet drops (2.4) is the same as that of Sect. 2.1. We assume that the packet loss process $\{\gamma_k\}$ is fed back to the sensor, which allows the sensor to reconstruct the error covariances $\{P_k\}$ at the remote estimator.

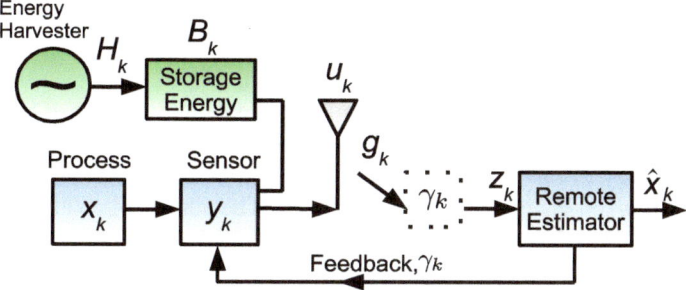

Fig. 2.4 Transmission power control with energy harvesting

In contrast to Sect. 2.1, here, the sensor is equipped with energy harvesting capabilities. Let the energy harvesting process be denoted by $\{H_k\}$, where H_k is the energy harvested between the discrete time instants $k-1$ and k. The process $\{H_k\}$ is modelled as a stationary, first-order, homogeneous Markov process, which is independent of the fading process $\{g_k\}$. This modelling for the harvested energy process is justified by empirical measurements in, e.g. the case of solar energy [38].

We assume that the dynamics of the stored battery energy $B_{(\cdot)}$ is given by the following first-order Markov model

$$B_{k+1} = \min\{B_k - u_k + H_{k+1}, B_{\max}\}, \quad k \geq 0, \tag{2.13}$$

where u_k is the transmission energy at time k, and $B_{\max}$ is the maximum stored energy in the battery.

2.2.1 Optimal Energy Allocation Problems

In this subsection, we formulate optimal transmission energy[5] allocation problems in order to minimize the trace of the receiver's expected estimation error covariance. Unlike the problem formulation in Sect. 2.1, here, the optimal energy policies are computed at the sensor, since the sensor has information about the energy harvesting and instantaneous battery levels, as well as knowledge of $\{P_k\}$ from the feedback of $\{\gamma_k\}$.

We consider the scenario of causal information, where the realizations of future wireless fading channel gains and harvested energies are not a priori known to the transmitter, see also Remark 2.4. More precisely, the information available at the sensor at any time $k \geq 1$ is given by

---

[5]We measure energy on a per channel use basis and we will refer to energy and power interchangeably in this chapter.

$$\mathbb{I}_k = \{s_t := (\gamma_{t-1}, g_t, H_t, B_t) : 1 \leq t \leq k\} \cup \mathbb{I}_0 \qquad (2.14)$$

where $\mathbb{I}_0 := \{g_0, H_0, B_0, P_0\}$ is the initial condition.

The information $\mathbb{I}_k$ is used at the sensor to decide u_k, the amount of transmission energy to use at time k. This quantity affects both the packet loss process and the amount of energy in the battery. A policy $\{u_k\}$ is feasible if the energy harvesting constraint $0 \leq u_k \leq B_k$ is satisfied for all $k \geq 1$. The admissible control set is then given by $\mathscr{U} := \{u_{(\cdot)} : u_k$ is adapted to sigma-field $\sigma(\mathbb{I}_k)$ and $0 \leq u_k \leq B_k\}$.

The optimization problems are now formulated as Markov decision processes for the following two cases:

(i) *Finite-time horizon*:

$$\min_{\{u_k : 0 \leq k \leq T-1\}} \sum_{k=0}^{T-1} \mathbb{E}[\mathrm{tr}(P_{k+1})] \qquad (2.15)$$

$$\text{s.t. } 0 \leq u_k \leq B_k \quad 0 \leq k \leq T-1$$

(ii) *Infinite-time horizon*:

$$\min_{\{u_k : k \geq 0\}} \limsup_{T \to \infty} \frac{1}{T} \sum_{k=0}^{T-1} \mathbb{E}[\mathrm{tr}(P_{k+1})] \qquad (2.16)$$

$$\text{s.t. } 0 \leq u_k \leq B_k \quad k \geq 0$$

where B_k is the stored battery energy available at time k, which satisfies the battery dynamics (2.13). It is evident that the transmission energy u_k at time k not only affects the amount of stored energy B_{k+1} available at time $k+1$, but thereby also the transmission energy u_{k+1}, since $0 \leq u_{k+1} \leq B_{k+1} = \min\{B_k - u_k + H_{k+1}, B_{\max}\}$ by (2.13). One of the key issues in solving problems (2.15) and (2.16) is to determine if one should use a lot of energy at time k, or save up some of the energy for use at future times.

We now give sufficient conditions under which the infinite horizon stochastic control problem (2.16) is well posed when the system matrix A is unstable. With well posed we, here, mean that an exponential boundedness condition for the expected estimation error covariance is satisfied. Let $\mathbb{G}$ and $\mathbb{H}$ be the time-invariant probability transition laws of the Markovian channel fading process $\{g_k\}$ and the Markovian harvested energy process $\{H_k\}$, respectively.

We introduce the following assumption:

Assumption 2.2.1 The channel fading process $\{g_k\}$, harvested energy process $\{H_k\}$ and the maximum battery storage $B_{\max}$ satisfy the following:

$$\sup_{(g,H)} \int_{g_k} \int_{H_k} (1 - h(g_k \min\{H_k, B_{\max}\})) \mathbb{P}(g_k|g_{k-1} = g) \mathbb{P}(H_k|H_{k-1} = H) dg_k d H_k$$

$$\leq \frac{r}{||A||^2}, \quad k \geq 0 \tag{2.17}$$

for some $r \in [0, 1)$, where $||A||$ denotes the spectral norm of A.

Theorem 2.2 *Assume that Assumption 2.2.1 holds. Then there exist energy allocations $\{u_k\}$ such that $\mathbb{E}[P_k]$ satisfies;*

$$\mathbb{E}[\mathrm{tr}(P_k)] \leq \alpha r^k + \beta, \quad k \geq 0 \tag{2.18}$$

for some nonnegative scalars α and β, and $r \in [0, 1)$. As a result, the stochastic optimal control problem (2.16) is well posed.

Proof Based on Theorem 1 of [39], a sufficient condition for exponential stability in the sense of (2.18) is that

$$\sup_{(g,H)} \mathbb{P}(\gamma_k = 0|g_{k-1} = g, H_{k-1} = H)$$

$$= \sup_{(g,H)} \int_{g_k} \int_{H_k} \mathbb{P}(\gamma_k = 0|g_k = g', H_k = H', g_{k-1} = g, H_{k-1} = H)$$

$$\times \mathbb{P}(g_k, H_k|g_{k-1} = g, H_{k-1} = H) dg_k d H_k$$

$$= \sup_{(g,H)} \int_{g_k} \int_{H_k} \mathbb{P}(\gamma_k = 0|g_k = g', H_k = H', g_{k-1} = g, H_{k-1} = H)$$

$$\times \mathbb{P}(g_k|g_{k-1} = g) \mathbb{P}(H_k|H_{k-1} = H) dg_k d H_k$$

$$= \sup_{(g,H)} \int_{g_k} \int_{H_k} (1 - h(g_k u_k)) \mathbb{P}(g_k|g_{k-1} = g) \mathbb{P}(H_k|H_{k-1} = H) dg_k d H_k \leq \frac{r}{||A||^2}$$

for some $r \in [0, 1)$. We now consider a suboptimal solution to the stochastic optimal control problem (2.16), where the full amount of energy harvested at each time step is used, i.e. $u_0 = B_0$ and $u_k = \min\{H_k, B_{\max}\}$ for $k \geq 1$. Then with this policy (2.17) will be a sufficient condition for (2.18) in terms of the channel fading process, harvested energy process and the maximum battery storage. Therefore, Assumption 2.2.1 provides a sufficient condition for the exponential boundedness (2.18) of the expected estimation error covariance. □

Remark 2.3 In general, condition (2.17) given by Assumption 2.2.1 may be difficult to verify for all values of g, H and k. However, if we assume that the channel fading and harvested energy processes are stationary, then it would not be necessary to verify the condition for all k. Furthermore, in the two most commonly used models, namely i.i.d. processes and finite state Markov chains, the condition (2.17) can be simplified as follows:

(i) If $\{g_k\}$ and $\{H_k\}$ are i.i.d., then (2.17) amounts to

$$\int_{g_k} \int_{H_k} (1 - h(g_k \min\{H_k, B_{\max}\})) \mathbb{P}(g_k) \mathbb{P}(H_k) dg_k dH_k \leq \frac{r}{||A||^2}.$$

(ii) If $\{g_k\}$ and $\{H_k\}$ are stationary finite state Markov chains with state spaces $\{1, \ldots, M\}$ and $\{1, \ldots, N\}$ respectively, then (2.17) becomes

$$\max_{(i,j)} \sum_{i'=1}^{M} \sum_{j'=1}^{N} (1 - h(i \min\{j, B_{\max}\})) \mathbb{P}(g_k = i' | g_{k-1} = i) \mathbb{P}(H_k = j' | H_{k-1} = j) \leq \frac{r}{||A||^2}.$$

2.2.2 Solutions to the Optimal Energy Allocation Problems

The stochastic control problems (2.15) and (2.16) can be regarded as constrained Markov decision process (MDP) [30] problems with $s_k := (P_k, g_k, H_k, B_k)$ as the state and u_k as the control action. We will approach the constrained MDPs (2.15) and (2.16) by the use of dynamic programming techniques.

Note that due to the existence of a perfect feedback link the sensor has knowledge about whether its transmissions have been received at the receiver or not. Hence, at time k the sensor knows $\{P_t : 0 \leq t \leq k\}$. The information available at the sensor at time instant $k \geq 0$ is given by (2.14), which can be easily shown to be equivalent to

$$\mathbb{I}_k := \{s_t = (P_t, g_t, H_t, B_t) : 0 \leq t \leq k\}.$$

The causal information $\mathbb{I}_k$ is used to decide the amount of transmit energy u_k to be used at time k. The transmit energy policy is computed offline using dynamic programming. We recall that a policy u_k is feasible if the energy harvesting constraints $0 \leq u_k \leq B_k = \min\{B_{k-1} - u_{k-1} + H_{k-1}, B_{\max}\}$ are satisfied for all $k \geq 1$.

For the finite-time horizon problem (2.15), we may define the value function at time k as

$$V_k(s) := \min_{\{u_l\}_{l=k}^{T-1}} \sum_{t=k}^{T-1} \mathbb{E}[\text{tr}(P_{t+1}) | s_t, u_t], \text{ s.t. } s_k = s.$$

The optimality equation or Bellman dynamic programming equation associated with the constrained stochastic control problem (2.15) is then given by

$$V_k(s_k) = \min_{0 \leq u_k \leq B_k} \left\{ \mathbb{E}[\text{tr}(P_{k+1}) | s_k, u_k] + \mathbb{E}[V_{k+1}(s_{k+1}) | s_k, u_k] \right\} \qquad (2.19)$$

with the terminal condition

$$V_T(s_T) := \min_{0 \leq u_T \leq B_T} \mathbb{E}[\text{tr}(P_{T+1}) | s_T, u_T] = \mathbb{E}[\text{tr}(P_{T+1}) | s_T, B_T],$$

where we use all available energy for transmission at the final time T.

The optimal transmission energy at time instant $k \geq 0$ is

$$u_k^*(s_k) = \arg \min_{0 \leq u_k \leq B_k} \left\{ \mathbb{E}[\mathrm{tr}(P_{k+1})|s_k, u_k] + \mathbb{E}[V_{k+1}(s_{k+1})|s_k, u_k] \right\} \tag{2.20}$$

where $V_{k+1}(\cdot)$ is the solution to the Bellman equation (2.19).

We now simplify the terms in (2.19). First, we have

$$\mathbb{E}[\mathrm{tr}(P_{k+1})|s_k, u_k] = \mathrm{tr}(AP_k A^T + Q) - f(g_k u_k)\mathrm{tr}\left(AP_k C^T[CP_k C^T + R]^{-1} CP_k A^T\right)$$

with the constraint that $0 \leq u_k \leq B_k$. On the other hand,

$$\mathbb{E}[V_{k+1}(s_{k+1})|s_k, u_k] = \int_{s_{k+1}} V_{k+1}(s_{k+1})F(ds_{k+1}|s_k, u_k)$$

$$= \int_{P_{k+1}, g_{k+1}, H_{k+1}, B_{k+1}} V_{k+1}(P_{k+1}, g_{k+1}, H_{k+1}, B_{k+1})F(d(P_{k+1}, g_{k+1}, H_{k+1}, B_{k+1})|P_k, g_k, H_k, B_k, u_k)$$

where F is the probability transition law. But this together with (2.13) implies that

$$\mathbb{E}[V_{k+1}(s_{k+1})|s_k, u_k]) = \int_{P_{k+1}, g_{k+1}, H_{k+1}} V_{k+1}\left(P_{k+1}, g_{k+1}, H_{k+1}, \min\{B_k - u_k + H_k, B_{\max}\}\right)$$

$$\times F(dP_{k+1}|P_k, g_k, u_k)\mathbb{G}(g_{k+1}|g_k)\mathbb{H}(H_{k+1}|H_k)$$

which follows from the fact that the mutually independent Markovian processes g_{k+1} and H_{k+1} are independent of P_{k+1}. This gives

$$\mathbb{E}[V_{k+1}(s_{k+1})|s_k, u_k]) \tag{2.21}$$

$$= \int_{g_{k+1}, H_{k+1}} \left\{ V_{k+1}\left(AP_k A^T + Q, g_{k+1}, H_{k+1}, \min\{B_k - u_k + H_k, B_{\max}\}\right) \right.$$

$$\times \left(1 - f(g_k u_k)\right)$$

$$+ V_{k+1}\left(AP_k A^T + Q - AP_k C^T[CP_k C^T + R]^{-1}CP_k A^T, g_{k+1}, H_{k+1}, \right.$$

$$\min\{B_k - u_k + H_k, B_{\max}\}\right)$$

$$\left. \times f(g_k u_k)\right\}\mathbb{G}(g_{k+1}|g_k)\mathbb{H}(H_{k+1}|H_k). \tag{2.22}$$

Define

$$\mathcal{L}(P, \gamma) \triangleq APA^T + Q - \gamma APC^T(CPC^T + R)^{-1}CPA^T \tag{2.23}$$

For the infinite-time horizon problem (2.16), we have the following:

Theorem 2.3 *Independent of the initial condition* $\mathbb{I}_0 = \{g_0, H_0, B_0, P_0\}$, *the value of the infinite-time horizon minimization problem (2.16) is given by* ρ, *which is the solution of the average cost optimality (Bellman) equation*

$$\rho + V(P, g, H, B) = \min_{0 \leq u \leq B} \Big\{ \mathbb{E}\big[\mathrm{tr}\big(\mathscr{L}(P, \gamma)\big)\big|P, g, u\big]$$
$$+ \mathbb{E}\Big[V\big(\mathscr{L}(P, \gamma), \tilde{g}, \tilde{H}, \min\{B - u + \tilde{H}, B_{max}\}\big)\big|P, g, H, u\Big]\Big\}, \quad (2.24)$$

where V *is the relative value function.*

Proof See Appendix. $\square$

We note that a discretized version of the Bellman equations (2.19) or (2.24) can be used for numerical computation to find solutions to the MDP problems (2.15) and (2.16).

Remark 2.4 The causal information pattern is clearly relevant to most practical scenarios. However, it is also instructive to consider the non-causal information scenario where the sensor has a priori information about the energy harvesting $\{H_k\}$ process and the fading channel gains $\{g_k\}$ for all time periods, including the future ones. This may be feasible in the situation of known environment where the wireless channel fading gains and the harvested energies are predictable with high accuracy [22]. Furthermore, the performance of the non-causal information case can serve as a benchmark (a lower bound) for the causal case. Indeed, we will present some performance comparisons between the causal and non-causal cases in Sect. 2.2.4. Note that the energy allocation problems for the non-causal case can be solved using similar techniques as in the current subsection, thus the details are omitted for brevity. $\square$

2.2.3 Structural Results on the Optimal Energy Allocation Policies

In this section, the structure of the optimal transmission energy policy (2.20) is studied for the case of the finite-time horizon stochastic control problem (2.15) with causal information. Following similar arguments, one can show similar structural results for the infinite-time horizon problem (2.16). We begin with a preliminary result, which will be needed for the proof of Theorem 2.4.

Lemma 2.3 *Suppose* $f(\cdot)$ *in (2.4) is a concave function in* u_k *given* g_k. *Then, for given* P_k, g_k *and* H_k, *the value function* $V_k(P_k, g_k, H_k, B_k)$ *in (2.19) is convex in* B_k *for* $0 \leq k \leq T$. *As a result,*

$$V_0(P_0, g_0, H_0, B_0) = \min_{\{0 \le u_k \le B_k\}_{k=0}^{T-1}} \sum_{k=0}^{T-1} \mathbb{E}[\text{tr}(P_{k+1})]$$

is convex in B_0.

Proof Recall that $s_k = (P_k, g_k, H_k, B_k)$. First, note that, for given P_T, g_T and H_T, the final time value function

$$V_T(s_T) = \min_{0 \le u_T \le B_T} \mathbb{E}[\text{tr}(P_{T+1})|s_T, u_T] = \mathbb{E}[\text{tr}(P_{T+1})|s_T, B_T]$$

is a convex function in B_T, due to the fact that $f(\cdot)$ is a concave function in u_k given g_k (see Lemma 2.2). Assume that $V_{k+1}(s_{k+1})$ is convex in B_{k+1} for given P_{k+1}, g_{k+1} and H_{k+1}. Then, for given H_k and u_k, the function

$$V_{k+1}(P_{k+1}, g_{k+1}, H_{k+1}, \min\{B_k - u_k + H_k, B_{\max}\})$$

is convex in B_k, since it is the minimum of the constant $V_{k+1}(P_{k+1}, g_{k+1}, H_{k+1}, B_{\max})$ and (by the induction hypothesis) the convex function $V_{k+1}(P_{k+1}, g_{k+1}, H_{k+1}, B_k - u_k + H_k)$. Since the expectation operator preserves convexity, $\mathbb{E}[V_{k+1}(s_{k+1})|s_k, u_k]$ given in (2.22) is a convex function in B_k. As $V_k(s_k)$ in (2.19) is the infimal convolution of two convex functions in B_k for given P_k, g_k and H_k, it is also convex in B_k (see the proof of Theorem 1 in [22]). □

The following result shows that for fixed P_k, g_k and H_k, the optimal energy allocated is increasing with the battery level.

Theorem 2.4 *Suppose $f(\cdot)$ in (2.4) is a concave function in u_k given g_k. Then, given P_k, g_k and H_k, the optimal energy policy $u_k^o(P_k, g_k, H_k, B_k)$ in (2.20) is non-decreasing in B_k for $0 \le k \le T$.*

Proof Assume P_k, g_k and H_k are fixed. Define

$$L(B, u) = \mathbb{E}[\text{tr}(P_{k+1})|P_k, g_k, u]$$
$$+ \mathbb{E}[V_{k+1}(P_{k+1}, g_{k+1}, H_{k+1}, \min\{B - u + H_k, B_{\max}\})|P_k, g_k, H_k, u].$$

We wish to show that $L(B, u)$ is submodular in (B, u), i.e. for every $u' \ge u$ and $B' \ge B$, we have [25]:

$$L(B', u') - L(B, u') \le L(B', u) - L(B, u). \tag{2.25}$$

It is evident that $\mathbb{E}[\text{tr}(P_{k+1})|P_k, g_k, u]$ is submodular in (B, u) since it is independent of B. Let

$$Z(x) := \mathbb{E}[V_{k+1}(P_{k+1}, g_{k+1}, H_{k+1}, \min\{x + H_k, B_{\max}\})|P_k, g_k, H_k, u].$$

Since $Z(x)$ is convex in x by Lemma 2.3, we have

$$Z(x + \varepsilon) - Z(x) \leq Z(y + \varepsilon) - Z(y), \quad x \leq y, \ \varepsilon \geq 0$$

(see Proposition 2.2.6 in [40]). Letting $x = B - u'$, $y = B - u$ and $\varepsilon = B' - B$, we then have the submodularity condition (2.25) for $\tilde{Z}(B, u) \triangleq Z(B - u)$ [22]. Therefore, $L(B, u)$ is submodular in (B, u). We then note that submodularity is a sufficient condition for optimality of monotone increasing policies [25], i.e. since $L(B, u)$ is submodular in (B, u), then $u^*(B) = \arg\min_u L(B, u)$ is non-decreasing in B. $\square$

As discussed in [22], the structural result of Theorem 2.4 implies that if u_k^{uc} is the unique solution to the convex unconstrained minimization problem

$$u_k^{\text{uc}}(P_k, g_k, H_k)$$
$$= \arg\min_{u_k} \left\{ \mathbb{E}[\text{tr}(P_{k+1})|P_k, g_k, u_k] + \mathbb{E}[V_{k+1}(P_{k+1}, g_{k+1}, H_{k+1})|P_k, g_k, H_k, u_k] \right\},$$

then the solution to the constrained problem (2.20), where $0 \leq u_k \leq B_k$, will be of the form

$$u_k^*(P_k, g_k, H_k, B_k) = \begin{cases} 0, & \text{if } u_k^{\text{uc}} \leq 0 \\ u_k^{\text{uc}}, & \text{if } 0 < u_k^{\text{uc}} < B_k \\ B_k, & \text{if } u_k^{\text{uc}} \geq B_k. \end{cases}$$

In the case that the transmission energy allocation u_k belongs to a two element set $\{E_0, E_1\}$, the monotonicity of Theorem 2.4 yields a threshold structure. This threshold structure implies that, for fixed P_k, g_k and H_k, the optimal transmission energy allocation is of the form

$$u_k^*(P_k, g_k, H_k, B_k) = \begin{cases} E_0, & \text{if } B_k \leq B^* \\ E_1, & \text{otherwise}, \end{cases}$$

where B^* is the corresponding battery storage threshold. The threshold structure of the optimal energy allocation policy in the case of a binary energy allocation set simplifies the implementation of the optimal energy allocation significantly. A stochastic gradient algorithm for computing B^* is presented in [41].

2.2.4 Numerical Studies

We present here numerical results for a scalar process with the following parameters: $A = 1.2$, $C = 1$, $Q = 1$, $R = 1$. We assume that the sensor uses a binary phase shift keying (BPSK) transmission scheme with b bits per packet. Therefore, (2.4) is of the form [26]:

Fig. 2.5 Infinite-time
horizon average error
covariance versus maximum
battery storage

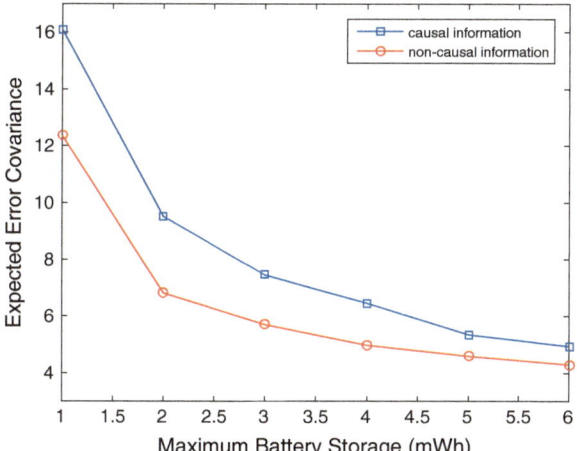

$$\mathbb{P}(\gamma_k = 1 | g_k, u_k) = f(g_k u_k) = \left(\int_{-\infty}^{\sqrt{g_k u_k}} \frac{1}{\sqrt{2\pi}} e^{-t^2/2} dt \right)^b$$

where we use $b = 4$ in the simulations.

The fading channel is taken to be Rayleigh [28], so that $\{g_k\}$ is i.i.d. exponentially distributed with probability density function (p.d.f) of the form $\mathbb{P}(g_k) = \frac{1}{\bar{g}} \exp(-g_k/\bar{g})$, with $\bar{g}$ being its mean. We also assume that the harvested energy process $\{H_k\}$ is i.i.d. and exponentially distributed, with p.d.f. $\mathbb{P}(H_k) = \frac{1}{\bar{H}} \exp(-H_k/\bar{H})$, with $\bar{H}$ being its mean.

For the following simulation results, we use 50 discretization points for each of the quantities P_k, g_k, B_k, u_k in the Bellman equations.

We first fix the mean of the fading channel gains to $\bar{g} = 1$ decibel (dB) and the mean of the harvested energy to $\bar{H} = 1$ milliwatt hour (mWh). Then, we plot in Fig. 2.5 the expected error covariance versus the maximum battery storage energy for the infinite-time horizon problem (2.16), where both cases of causal and non-causal fading channel gains and energy harvesting information are shown, see Remark 2.4. We see that the performance gets better as the maximum battery storage energy increases in both cases. Figure 2.5 also shows that, as expected, the performance for the non-causal information case is generally better than the performance of the system with only causal information.

Finally, we fix the mean of the harvested energy to $\bar{H} = 1$ (mWh), and the maximum battery storage energy to 2 (mWh). For the infinite-time horizon formulation (2.16), the expected error covariance versus the mean of the fading channel gains is plotted in Fig. 2.6, for both cases of causal and non-causal information. As shown in Fig. 2.6, in both cases, the performance improves as the mean of the fading channel gain increases.

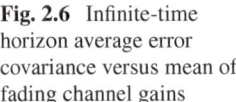

Fig. 2.6 Infinite-time
horizon average error
covariance versus mean of
fading channel gains

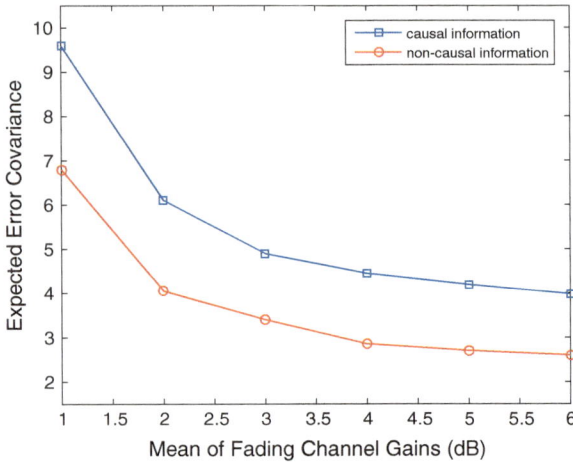

2.3 Conclusion

In this chapter, we have investigated transmission power control for Kalman filtering
with random packet drops over a fading channel, where the packet reception probability depends on both the time-varying fading channel gain and the sensor transmit
power. We first studied the problem of minimizing the trace of the expected error
covariance subject to an average power constraint. The resulting Markov decision
process problems are solved by the use of dynamic programming techniques. Simpler
suboptimal power allocation policies such as a constant power allocation policy and
a truncated channel inversion policy have also been considered. Numerical studies
suggest that, for low average transmit powers, optimal power allocation significantly
outperforms the suboptimal policies, while for higher average transmit powers, the
performance of the truncated channel inversion policy approaches the performance
of the optimal policy.

We then studied the problem of optimal transmission energy allocation for estimation error covariance minimization, when the sensor is equipped with energy harvesting capabilities. In this problem formulation, the trace of the expected estimation
error covariance of the Kalman filter is minimized, subject to energy harvesting constraints. Using the concept of submodularity, some structural results on the optimal
transmission energy allocation policy have also been obtained.

Notes: Section 2.1 is based on [42], while Sect. 2.2 is based on [41]. The case
of imperfect feedback acknowledgements, and a stochastic gradient algorithm for
computing the threshold in the case of binary energy levels, is also considered in
[41]. The work of [41] has since been extended to control with an energy harvesting
in [43]. Energy harvesting in the context of estimation and control has also been
subsequently studied in [44, 45], see also Sect. 3.2.

In this book, power allocation decisions are often made at the remote estimator (this is analogous to the situation in wireless communications, where power allocation is often done at the base station and fed back to the mobiles), which motivates us to consider decisions based on the estimation error covariance. When power allocation decisions are made at the sensor, researchers have tried to make use of additional state (or measurement) information [46, 47].

Appendix

Proof of Lemma 2.2

Proof The proof uses similar ideas to the proof of Proposition 3.1 in [36]. The decreasing property follows from the relation

$$\mathbb{E}[P_{k+1}] = \mathbb{E}[P_{k+1}|P_k, g_k, u_k]$$
$$= \mathbb{E}[AP_kA^T + Q - f(g_ku_k)AP_kC^T(CP_kC^T + R)^{-1}CP_kA^T]$$

and the assumption that $f(.)$ is an increasing function.

For the proof of convexity, let u^1 and u^2 be two average transmit powers, where $u^1 \neq u^2$, with $\mathfrak{p}^*(u^1)$ and $\mathfrak{p}^*(u^2)$ the corresponding traces of the expected error covariances. We want to show that

$$\mathfrak{p}^*(\lambda u^1 + (1 - \lambda)u^2) < \lambda \mathfrak{p}^*(u^1) + (1 - \lambda)\mathfrak{p}^*(u^2), \forall \lambda \in (0, 1).$$

Let $\{u_k^1(P_k, g_k)\}$ be the optimal power allocation policy that achieves $\mathfrak{p}^*(u^1)$, and $\{u_k^2(P_k, g_k)\}$ be the optimal power allocation policy that achieves $\mathfrak{p}^*(u^2)$. Define a new policy $\{u_k^\lambda(P_k, g_k)\}$ such that

$$u_k^\lambda(P_k, g_k) = \lambda u_k^1(P_k, g_k) + (1 - \lambda)u_k^2(P_k, g_k), \forall P_k, g_k.$$

We will first show that for a given P_k, we have:

(1) $\mathbb{E}[u_k^\lambda|P_k] \leq \lambda \mathbb{E}[u_k^1|P_k] + (1 - \lambda)\mathbb{E}[u_k^2|P_k]$, and
(2) $\mathbb{E}[\text{tr}(P_{k+1}^\lambda)|P_k] < \lambda \mathbb{E}[\text{tr}(P_{k+1}^1)|P_k] + (1 - \lambda)\mathbb{E}[\text{tr}(P_{k+1}^2)|P_k]$,

where P_{k+1}^j is the value of P_{k+1} that follows from using policy $\{u_k^j(.)\}$, for $j = 1, 2, \lambda$, respectively. For (1), this clearly follows from the definition of u_k^λ. For (2), we have

$$\mathbb{E}[\mathrm{tr}(P_{k+1}^{\lambda})|P_k]$$

$$= \int \left(\mathrm{tr}(AP_kA^T + Q) - f(g_ku_k^{\lambda})\mathrm{tr}(AP_kC^T(CP_kC^T + R)^{-1}CP_kA^T) \right) F(dg_k)$$

$$< \int \left(\mathrm{tr}(AP_kA^T + Q) - (\lambda f(g_ku_k^1) + (1 - \lambda)f(g_ku_k^2)) \right.$$

$$\left. \times \mathrm{tr}(AP_kC^T(CP_kC^T + R)^{-1}CP_kA^T) \right) F(dg_k)$$

$$= \lambda\mathbb{E}[\mathrm{tr}(P_{k+1}^1)|P_k] + (1 - \lambda)\mathbb{E}[\mathrm{tr}(P_{k+1}^2)|P_k]$$

where the inequality comes from the strict concavity of $f(.)$.

From (1) and (2), we have

$$\lim_{K\to\infty} \frac{1}{K}\sum_{k=1}^{K} \mathbb{E}[u_k^{\lambda}] = \lim_{K\to\infty} \frac{1}{K}\sum_{k=1}^{K} \mathbb{E}[\mathbb{E}[u_k^{\lambda}|P_k]]$$

$$\leq \lim_{K\to\infty} \frac{1}{K}\sum_{k=1}^{K} \mathbb{E}[\lambda\mathbb{E}[u_k^1|P_k] + (1 - \lambda)\mathbb{E}[u_k^2|P_k]]$$

$$= \lambda u^1 + (1 - \lambda)u^2$$

and

$$\lim_{K\to\infty} \frac{1}{K}\sum_{k=1}^{K} \mathbb{E}[\mathrm{tr}(P_{k+1}^{\lambda})] = \lim_{K\to\infty} \frac{1}{K}\sum_{k=1}^{K} \mathbb{E}[\mathbb{E}[\mathrm{tr}(P_{k+1}^{\lambda})|P_k]]$$

$$< \lim_{K\to\infty} \frac{1}{K}\sum_{k=1}^{K} \mathbb{E}\left[\lambda\mathbb{E}[\mathrm{tr}(P_{k+1}^1)|P_k] + (1 - \lambda)\mathbb{E}[\mathrm{tr}(P_{k+1}^2)|P_k] \right]$$

$$= \lambda\mathfrak{p}^*(u^1) + (1 - \lambda)\mathfrak{p}^*(u^2).$$

By the definition of $\mathfrak{p}^*(u)$ being the minimum expected error covariance such that the average transmit power is less than or equal to u, we then have $\mathfrak{p}^*(\lambda u^1 + (1-\lambda)u^2) \leq \frac{1}{K}\sum_{k=1}^{K} \mathbb{E}[\mathrm{tr}(P_{k+1}^{\lambda})] < \lambda\mathfrak{p}^*(u^1) + (1 - \lambda)\mathfrak{p}^*(u^2)$.

Proof of Theorem 2.3

We first establish the inequality

$$\rho + V(P, g, H, B) \geq \min_{0 \leq u \leq B} \left\{ \mathbb{E}[\mathrm{tr}(\mathscr{L}(P, \gamma))|P, g, u] \right.$$

$$\left. + \mathbb{E}\left[V(\mathscr{L}(P, \gamma), \tilde{g}, \tilde{H}, \min\{B - u + \tilde{H}, B_{\max}\})|P, g, H, u \right] \right\} \qquad (2.26)$$

by verifying conditions (W) and (B) of [48], that guarantee the existence of solutions to (2.26) for MDPs with general state space. Denote the state space by $\mathscr{S}$ and action space by $\mathscr{A}$, i.e. $(P_k, g_k, H_k, B_k) \in \mathscr{S}$ and $u_k \in \mathscr{A}$. Condition (W) of [48] in our notation says that:

(1) The state space $\mathscr{S}$ is locally compact.
(2) Let $U(\cdot)$ be the mapping that assigns to each (P_k, g_k, H_k, B_k) the nonempty set of available actions. Then $U(P_k, g_k, H_k, B_k)$ lies in a compact subset of $\mathscr{A}$ and $U(\cdot)$ is upper semicontinuous.
(3) The transition probabilities are weakly continuous.
(4) $\mathbb{E}[\mathrm{tr}(\mathscr{L}(P, \gamma)) | P, g, u]$ is lower semicontinuous.
By our assumption that $u_k \leq B_k \leq B_{\max}$, (0) and (1) of (W) can be easily verified. The conditions (2) and (3) follow from the definition (2.23).

Define $w_\delta(P_0, g_0, H_0, B_0) = v_\delta(P_0, g_0, H_0, B_0) - m_\delta$, where

$$v_\delta(P_0, g_0, H_0, B_0) = \inf_{\{u_k : k \geq 0\}} \mathbb{E}\left[\sum_{k=0}^{\infty} \delta^k \mathbb{E}[\mathrm{tr}(\mathscr{L}(P_k, \gamma_k)) | P_k, g_k, u_k] \Big| P_0, g_0, H_0, B_0 \right]$$

and $m_\delta = \inf_{(P_0, g_0, H_0, B_0)} v_\delta(P_0, g_0, H_0, B_0)$. Condition (B) of [48] in our notation says that

$$\sup_{\delta < 1} w_\delta(P_0, g_0, H_0, B_0) < \infty, \qquad \forall (P_0, g_0, H_0, B_0).$$

Following Sect. 4 of [48], define the stopping time

$$\tau = \inf\{k \geq 0 : v_\delta(P_k, g_k, H_k, B_k) \leq m_\delta + \varsigma\}$$

for some $\varsigma \geq 0$. Given $\varsigma > 0$ and an arbitrary (P_0, g_0, H_0, B_0), consider a suboptimal power allocation policy where the sensor transmits based on the same policy as the one that achieves m_δ (with a different initial condition) until $v_\delta(P_N, g_N, H_N, B_N) \leq m_\delta + \varsigma$ is satisfied at some time N. By the exponential forgetting property of initial conditions for Kalman filtering, we have $N < \infty$ with probability 1 and $\mathbb{E}[N] < \infty$. Since $\tau \leq N$, we have $\mathbb{E}[\tau] < \infty$. Then by Lemma 4.1 of [48],

$$w_\delta(P_0, g_0, H_0, B_0) \leq \varsigma + \inf_{\{\gamma_k\}} \mathbb{E}\left[\sum_{k=0}^{\tau-1} \mathbb{E}[\mathrm{tr}(\mathscr{L}(P_k, \gamma_k)) | P_k, g_k, u_k] \Big| P_0, g_0, H_0, B_0 \right]$$

$$\leq \varsigma + \mathbb{E}[\tau] \times Z < \infty \tag{2.27}$$

where the second inequality uses Wald's equation, with Z being an upper bound to the expected error covariance, which exists by Theorem 2.2. Hence, condition (B) of [48] is satisfied and a solution to the inequality (2.26) exists.

To show equality in (2.26), we will require a further equicontinuity property of the optimal cost for the related discounted cost MDP to be satisfied. This can be shown by a similar argument as in the proof of Proposition 3.2 of [49]. The assumptions in Sects. 5.4 and 5.5 of [33] may then be verified to conclude the existence of a solution to the average cost optimality equation (2.24).

References

1. B. Sinopoli, L. Schenato, M. Franceschetti, K. Poolla, M.I. Jordan, S.S. Sastry, Kalman filtering with intermittent observations. IEEE Trans. Autom. Control **49**(9), 1453–1464 (2004)
2. X. Liu, A.J. Goldsmith, Kalman filtering with partial observation losses, in *Proceedings of the IEEE Conference on Decision and Control*, Bahamas (2004), pp. 1413–1418
3. V. Gupta, N.C. Martins, J.S. Baras, Optimal output feedback control using two remote sensors over erasure channels. IEEE Trans. Autom. Control **54**(7), 1463–1476 (2009)
4. K. Plarre, F. Bullo, On Kalman filtering for detectable systems with intermittent observations. IEEE Trans. Autom. Control **54**(2), 386–390 (2009)
5. Y. Mo, B. Sinopoli, Kalman filtering with intermittent observations: tail distribution and critical value. IEEE Trans. Autom. Control **57**(3), 677–689 (2012)
6. M. Epstein, L. Shi, A. Tiwari, R.M. Murray, Probabilistic performance of state estimation across a lossy network. Automatica **44**, 3046–3053 (2008)
7. L. Shi, M. Epstein, R.M. Murray, Kalman filtering over a packet-dropping network: a probabilistic perspective. IEEE Trans. Autom. Control **55**(3), 594–604 (2010)
8. Y. Xu, J.P. Hespanha, Estimation under uncontrolled and controlled communications in networked control systems, in *Proceedings of the IEEE Conference on Decision and Control*, Seville, Spain (2005), pp. 842–847
9. L. Schenato, Optimal estimation in networked control systems subject to random delay and packet drop. IEEE Trans. Autom. Control **53**(5), 1311–1317 (2008)
10. M. Huang, S. Dey, Stability of Kalman filtering with Markovian packet losses. Automatica **43**, 598–607 (2007)
11. K. You, M. Fu, L. Xie, Mean square stability for Kalman filtering with Markovian packet losses. Automatica **47**(12), 1247–1257 (2011)
12. A.F. Molisch, *Wireless Communications*, 2nd edn. (Wiley, New York, 2011)
13. D.N.C. Tse, P. Viswanath, *Fundamentals of Wireless Communication* (Cambridge University Press, Cambridge, 2005)
14. A.S. Leong, S. Dey, G.N. Nair, P. Sharma, Power allocation for outage minimization in state estimation over fading channels. IEEE Trans. Signal Process. **59**(7), 3382–3397 (2011)
15. D.E. Quevedo, A. Ahlén, J. Østergaard, Energy efficient state estimation with wireless sensors through the use of predictive power control and coding. IEEE Trans. Signal Process. **58**(9), 4811–4823 (2010)
16. D.E. Quevedo, A. Ahlén, A.S. Leong, S. Dey, On Kalman filtering over fading wireless channels with controlled transmission powers. Automatica **48**(7), 1306–1316 (2012)
17. M.M. Tentzeris, A. Georgiadis, L. Roselli (eds.), Special issue on energy harvesting and scavenging. Proc. IEEE **102**(11) (2014)
18. S. Ulukus, E. Erkip, P. Grover, K. Huang, O. Simeone, A. Yener, M. Zorzi (eds.), Special issue on wireless communications powered by energy harvesting and wireless energy transfer. IEEE J. Sel. Areas Commun. **33**(3) (2015)
19. D. Niyato, E. Hossain, M. Rashid, V. Bhargava, Wireless sensor networks with energy harvesting technologies: a game-theoretic approach to optimal energy management. IEEE Trans. Wirel. Commun. **14**(4), 90–96 (2007)
20. V. Sharma, U. Mukherji, V. Joseph, S. Gupta, Optimal energy management policies for energy harvesting sensor nodes. IEEE Trans. Wirel. Commun. **9**(4), 1326–1336 (2010)

21. O. Ozel, K. Tutuncuoglu, J. Yang, S. Ulukus, A. Yener, Transmission with energy harvesting nodes in fading wireless channels: optimal policies. IEEE J. Sel. Areas Commun. **29**(8), 1732–1743 (2011)
22. C.K. Ho, R. Zhang, Optimal energy allocation for wireless communications with energy harvesting constraints. IEEE Trans. Signal Process. **60**(9), 4808–4818 (2012)
23. M. Kashef, A. Ephremides, Optimal packet scheduling for energy harvesting sources on time varying wireless channels. J. Commun. Netw. **14**(2), 121–129 (2012)
24. A. Nayyar, T. Başar, D. Teneketzis, V.V. Veeravalli, Optimal strategies for communication and remote estimation with an energy harvesting sensor. IEEE Trans. Autom. Control **58**(9), 2246–2260 (2013)
25. D.M. Topkis, *Supermodularity and Complementarity* (Princeton University Press, Princeton, 2001)
26. J.G. Proakis, *Digital Communications*, 4th edn. (McGraw-Hill, New York, 2001)
27. G. Caire, G. Taricco, E. Biglieri, Optimum power control over fading channels. IEEE Trans. Inf. Theory **45**(5), 1468–1489 (1999)
28. T.S. Rappaport, *Wireless Communications: Principles and Practice*, 2nd edn. (Prentice Hall, Upper Saddle River, 2002)
29. D.E. Quevedo, J. Østergaard, A. Ahlén, Power control and coding formulation for state estimation with wireless sensors. IEEE Trans. Control Syst. Technol. **22**(2), 413–427 (2014)
30. O. Hernández-Lerma, J. González-Hernández, R.R. López-Martínez, Constrained average cost Markov control processes in Borel spaces. SIAM J. Control Optim. **42**(2), 442–468 (2003)
31. N. Ghasemi, S. Dey, Power-efficient dynamic quantization for multisensor HMM state estimation over fading channels, in *Proceedings of the ISCCSP*, Malta (2008), pp. 1553–1558
32. D.P. Bertsekas, *Dynamic Programming and Optimal Control*, vol. I, 2nd edn. (Athena Scientific, Belmont, 2000)
33. O. Hernández-Lerma, J.B. Lasserre, *Discrete-Time Markov Control Processes: Basic Optimality Criteria* (Springer, Berlin, 1996)
34. H. Yu, D.P. Bertsekas, Discretized approximations for POMDP with average cost, in *Proceedings of the 20th Conference on Uncertainty in Artifical Intelligence*, Banff, Canada (2004), pp. 619–627
35. S. Boyd, L. Vandenberghe, *Convex Optimization* (Cambridge University Press, Cambridge, 2004)
36. R.A. Berry, R.G. Gallager, Communication over fading channels with delay constraints. IEEE Trans. Inf. Theory **48**(5), 1135–1149 (2002)
37. A.J. Goldsmith, P.P. Varaiya, Capacity of fading channels with channel side information. IEEE Trans. Inf. Theory **43**(6), 1986–1992 (1997)
38. C.K. Ho, P.D. Khoa, P.C. Ming, Markovian models for harvested energy in wireless communications," in *IEEE International Conference on Communication System (ICCS)*, Singapore (2010), pp. 311–315
39. D.E. Quevedo, A. Ahlén, K.H. Johansson, State estimation over sensor networks with correlated wireless fading channels. IEEE Trans. Autom. Control **58**(3), 581–593 (2013)
40. D. Simchi-Levi, X. Chen, J. Bramel, *The Logic of Logistics: Theory, Algorithms, and Applications for Logistics and Supply Chain Management* (Springer, Berlin, 2004)
41. M. Nourian, A.S. Leong, S. Dey, Optimal energy allocation for Kalman filtering over packet dropping links with imperfect acknowledgments and energy harvesting constraints. IEEE Trans. Autom. Control **59**(8), 2128–2143 (2014)
42. A.S. Leong, S. Dey, Power allocation for error covariance minimization in Kalman filtering over packet dropping links, in *Proceedings of the IEEE Conference on Decision and Control*, Maui, HI (2012), pp. 3335–3340
43. S. Knorn, S. Dey, Optimal energy allocation for linear control with packet loss under energy harvesting constraints. Automatica **77**, 259–267 (2017)
44. Y. Li, F. Zhang, D.E. Quevedo, V. Lau, S. Dey, L. Shi, Power control of an energy harvesting sensor for remote state estimation. IEEE Trans. Autom. Control **62**(1), 277–290 (2017)

45. J. Huang, D. Shi, T. Chen, Event-triggered state estimation with an energy harvesting sensor. IEEE Trans. Autom. Control (2017), to appear
46. J. Wu, Y. Li, D.E. Quevedo, V. Lau, L. Shi, Data-driven power control for state estimation: a Bayesian inference approach. Automatica **54**, 332–339 (2015)
47. K. Gatsis, A. Ribeiro, G.J. Pappas, Optimal power management in wireless control systems. IEEE Trans. Autom. Control **59**(6), 1495–1510 (2014)
48. M. Schäl, Average optimality in dynamic programming with general state space. Math. Oper. Res. **18**(1), 163–172 (1993)
49. M. Huang, S. Dey, Dynamic quantizer design for hidden Markov state estimation via multiple sensors with fusion center feedback. IEEE Trans. Signal Process. **54**(8), 2887–2896 (2006)

Chapter 3
Optimal Transmission Scheduling for Event-Triggered Estimation with Packet Drops

As we have seen in the previous chapters, energy usage is a crucial issue in wireless networked control systems. The concept of event-triggered estimation of dynamical systems, where sensor measurements or state estimates are sent to a remote estimator/controller only when certain events occur, has gained significant recent attention. By transmitting only when necessary, as dictated by performance objectives, e.g. such as when the estimation quality at the remote estimator has deteriorated sufficiently, potential savings in energy usage, which are important in networked estimation and control applications, can be achieved.

Event-triggered estimation has been investigated in e.g. [1–11], while event-triggered control has been studied in e.g. [12–16]. Many rules for deciding when a sensor should transmit have been proposed in the literature, such as if the estimation error [2, 4, 5, 8], error in predicted output [11], other functions of the estimation error [3, 9, 10], or the error covariance [7], exceed a given threshold. These transmission policies often lead to energy savings. However, the motivation for using these rules is usually based on heuristics. Furthermore, mostly the idealized case, where all transmissions (when scheduled) are received at the remote estimator, is considered. Packet drops [17], which are unavoidable when using a wireless communication medium, are neglected in these works, saved for some works in event-triggered control [14, 16].

In Sect. 3.1, we consider an event-triggered estimation problem with i.i.d. packet drops, and derive structural properties on the optimal transmission schedule.[1] For transmission schedules which decide whether to transmit local state estimates based

---

[1] Optimal transmission scheduling can in some ways be regarded as less general than the optimal power allocation problems studied in Chap. 2. On the other hand, stronger structural results can be obtained when we restrict ourselves to transmission scheduling problems where the possible sensor decisions are either transmit or don't transmit.

© The Author(s) 2018

A.S. Leong et al., *Optimal Control of Energy Resources for State Estimation Over Wireless Channels*, SpringerBriefs in Control, Automation and Robotics, DOI 10.1007/978-3-319-65614-4_3

only on knowledge of the error covariance at the remote estimator, our analysis shows that a *threshold policy*, where the sensor transmits if the error covariance exceeds a threshold and does not transmit otherwise, is indeed optimal. For noiseless measurements and no packet drops, similar structural results were derived using majorization theory for scalar [18] and vector [19] systems respectively. In the situation where sensor measurements (rather than local estimates) are transmitted, the optimality of threshold policies is proved for the scalar case in Sect. 3.1.4. These structural results provide a theoretical justification for the use of such variance-based threshold policies in event-triggered estimation. However, for vector systems, we provide a counterexample to show that, in general, threshold-type policies are not optimal when measurements are transmitted.

In Sect. 3.2, we study the case where the sensor is equipped with energy harvesting capabilities (see Chap. 2, [20, 21], and the references therein), and transmission over a packet dropping channel can only occur if there is sufficient energy in the battery. We will prove that for a given battery energy level and a given harvested energy, the optimal policy is a threshold policy on the error covariance. Similarly, for a given error covariance and a given harvested energy, the optimal policy is a threshold policy on the battery level.

3.1 Transmission Scheduling over a Packet Dropping Channel

3.1.1 System Model

A diagram of the system model for this section is shown in Fig. 3.1. Consider a discrete-time process

$$x_{k+1} = Ax_k + w_k, \tag{3.1}$$

where $x_k \in \mathbb{R}^n$ and w_k is i.i.d. Gaussian with zero mean and covariance $Q > 0$.[2] There is a sensor taking measurements

$$y_k = Cx_k + v_k, \tag{3.2}$$

where $y_k \in \mathbb{R}^m$ and v_k is i.i.d. Gaussian with zero mean and covariance $R > 0$. We assume that $\{w_k\}$ and $\{v_k\}$ are mutually independent, the pair (A, C) is detectable and the pair $(A, Q^{1/2})$ is stabilizable.

In contrast to the situation examined in Chap. 2, here, the sensor has computational capability that allows it to run a local Kalman filter. The state estimates and error covariances at the sensor, namely

---

[2]We say that a matrix $X > 0$ if X is positive definite, and $X \geq 0$ if X is positive semi-definite.

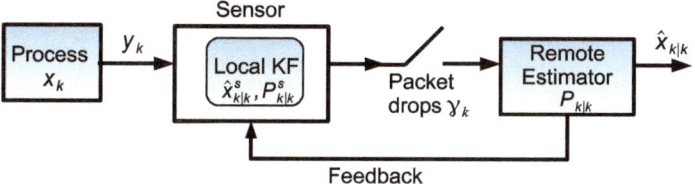

Fig. 3.1 Transmission scheduling for remote state estimation

$$\hat{x}_{k|k-1}^s \triangleq \mathbb{E}[x_k|y_0, \ldots, y_{k-1}]$$

$$\hat{x}_{k|k}^s \triangleq \mathbb{E}[x_k|y_0, \ldots, y_k]$$

$$P_{k|k-1}^s \triangleq \mathbb{E}[(x_k - \hat{x}_{k|k-1}^s)(x_k - \hat{x}_{k|k-1}^s)^T|y_0, \ldots, y_{k-1}]$$

$$P_{k|k}^s \triangleq \mathbb{E}[(x_k - \hat{x}_{k|k}^s)(x_k - \hat{x}_{k|k}^s)^T|y_0, \ldots, y_k],$$

can be computed using the Kalman filter equations. Let $\bar{P}^s$ be the steady state value of $P_{k|k-1}^s$ and $\bar{P}$ the steady state value of $P_{k|k}^s$ as $k \to \infty$, which both exist due to the detectability assumption. To simplify the presentation, we will assume that the local Kalman filter is operating in steady state, so that $P_{k|k-1}^s = \bar{P}^s$ and $P_{k|k}^s = \bar{P}, \forall k$. Note that in general the local Kalman filter will converge to steady state at an exponential rate [22].

At time k, the remote estimator decides whether or not the sensor should send its current state estimate $\hat{x}_{k|k}^s$. Let $\nu_k \in \{0, 1\}$ be decision variables such that $\nu_k = 1$ if $\hat{x}_{k|k}^s$ is transmitted to the remote estimator, and $\nu_k = 0$ if there is no transmission. We will assume that ν_k is computed by the remote estimator based on information available at time $k - 1$, and fed back to the sensor via a feedback link, see Fig. 3.1.[3] The decision ν_k is assumed to not depend on the current value of x_k (or functions of x_k such as measurements and state estimates). In particular, we will assume that ν_k depends only on the error covariances at the remote estimator, see Sect. 3.1.2.

Sensor transmissions occur over a packet dropping link. Let γ_k be random variables such that $\gamma_k = 1$ if the transmission at time k is successfully received by the remote estimator, and $\gamma_k = 0$ otherwise. We begin our analysis by assuming that $\{\gamma_k\}$ is i.i.d. Bernoulli with

$$\mathbb{P}(\gamma_k = 1) = \lambda, \quad \lambda \in (0, 1).$$

Define the information set available to the remote estimator at time k as:

$$\mathbb{I}_k \triangleq \{\nu_0, \ldots, \nu_k, \nu_0\gamma_0, \ldots, \nu_k\gamma_k, \nu_0\gamma_0\hat{x}_{0|0}^s, \ldots, \nu_k\gamma_k\hat{x}_{k|k}^s\}.$$

Denote the state estimates and error covariances at the remote estimator by:

---

[3]Scheduling can also be done at the sensor if γ_{k-1} is fed back from the remote estimator to the sensor.

$$\hat{x}_{k|k} \triangleq \mathbb{E}[x_k | \mathbb{I}_k]$$
$$P_{k|k} \triangleq \mathbb{E}[(x_k - \hat{x}_{k|k})(x_k - \hat{x}_{k|k})^T | \mathbb{I}_k].$$

Given that the decision variables v_k depend on $P_{k-1|k-1}$, but not on the state x_k (or functions of x_k), the optimal remote estimator can be shown to have the following form, similar to [23, 24]:

$$
\begin{aligned}
\hat{x}_{k|k} &= \begin{cases} \hat{x}_{k|k}^s & , \text{ if } v_k \gamma_k = 1 \\ A\hat{x}_{k-1|k-1} & , \text{ if } v_k \gamma_k = 0 \end{cases} \\
P_{k|k} &= \begin{cases} \bar{P} & , \text{ if } v_k \gamma_k = 1 \\ A P_{k-1|k-1} A^T + Q & , \text{ if } v_k \gamma_k = 0. \end{cases}
\end{aligned}
\tag{3.3}
$$

For our subsequent analysis, we introduce

$$f(X) \triangleq AXA^T + Q \tag{3.4}$$

and define the countably infinite set

$$\mathbb{S} \triangleq \{\bar{P}, f(\bar{P}), f^2(\bar{P}), \ldots\}, \tag{3.5}$$

where $f^n(.)$ denotes the n-fold composition of $f(.)$, with the convention that $f^0(X) = X$. Then, it is clear from (3.3) that $\mathbb{S}$ consists of all possible values of $P_{k|k}$ at the remote estimator.

3.1.2 Optimization of Transmission Scheduling

As stated in the previous subsection, we will consider transmission policies where $v_k(P_{k-1|k-1})$ depends only on $P_{k-1|k-1}$, similar to [7]. To take into account energy usage, we will assume a fixed transmission energy cost of E for each scheduled transmission (when $v_k = 1$, independent of the reception outcome γ_k). We will consider the following finite horizon (of horizon K) optimization problem:

$$\min_{\{v_k\}} \sum_{k=1}^{K} \mathbb{E}\left[\beta \operatorname{tr} P_{k|k} + (1 - \beta) v_k E\right], \tag{3.6}$$

for some design parameter $\beta \in (0, 1)$. Problem (3.6) minimizes a linear combination of the trace of the expected error covariance at the remote estimator and the expected transmission energy. We can write

$$\min_{\{v_k\}} \sum_{k=1}^{K} \mathbb{E}\left[\beta \mathrm{tr} P_{k|k} + (1-\beta)v_k E\right] = \min_{\{v_k\}} \sum_{k=1}^{K} \mathbb{E}\left[\mathbb{E}\left[\beta \mathrm{tr} P_{k|k} + (1-\beta)v_k E \mid P_{0|0}, \mathbb{I}_{k-1}, v_k\right]\right]$$

$$= \min_{\{v_k\}} \sum_{k=1}^{K} \mathbb{E}\left[\mathbb{E}\left[\beta \mathrm{tr} P_{k|k} + (1-\beta)v_k E \mid P_{k-1|k-1}, v_k\right]\right],$$

where the last line holds since $P_{k-1|k-1}$ is a deterministic function of $P_{0|0}$ and $\mathbb{I}_{k-1}$, and $P_{k|k}$ is a function of $P_{k-1|k-1}$, v_k, and γ_k. We note that

$$\mathbb{E}[\mathrm{tr} P_{k|k} \mid P_{k-1|k-1}, v_k] = v_k \left[\lambda \mathrm{tr} \bar{P} + (1-\lambda)\mathrm{tr} f(P_{k-1|k-1})\right] + (1-v_k)\mathrm{tr} f(P_{k-1|k-1})$$

$$= v_k \lambda \mathrm{tr} \bar{P} + (1-v_k\lambda)\mathrm{tr} f(P_{k-1|k-1})$$

(3.7)

where $f(.)$ is defined in (3.4). Let the functions $J_k(.) : \mathbb{S} \to \mathbb{R}$ be defined recursively as:

$$J_{K+1}(P) = 0$$

$$J_k(P) = \min_{v \in \{0,1\}} \left\{ \beta \left[v\lambda \mathrm{tr} \bar{P} + (1-v\lambda)\mathrm{tr} f(P)\right] + (1-\beta)v E + v\lambda J_{k+1}(\bar{P}) \right.$$

$$\left. + (1-v\lambda) J_{k+1}(f(P)) \right\}, \quad k = K, \ldots, 1.$$

(3.8)

Problem (3.6) can then be solved using the dynamic programming algorithm by computing $J_k(P_{k-1|k-1})$ for $k = K, K-1, \ldots, 1$, with the optimal $v_k^* = \mathrm{argmin}\, J_k(P_{k-1|k-1})$. Note that the finite horizon problem (3.6) can be solved exactly via explicit enumeration, since for a given initial $P_{0|0}$, the number of possible values for $P_{k|k}, k = 1, \ldots, K$ is finite.

We will also consider the infinite horizon problem:

$$\min_{\{v_k\}} \limsup_{K \to \infty} \frac{1}{K} \sum_{k=1}^{K} \mathbb{E}\left[\mathbb{E}\left[\beta \mathrm{tr} P_{k|k} + (1-\beta)v_k E \mid P_{k-1|k-1}, v_k\right]\right], \quad (3.9)$$

which is a Markov decision process (MDP) based stochastic control problem with v_k as the 'action' and $P_{k-1|k-1}$ as the 'state' at time k. The Bellman equation for problem (3.9) is

$$\rho + h(P) = \min_{v \in \{0,1\}} \left\{ \beta \left[v\lambda \mathrm{tr} \bar{P} + (1-v\lambda)\mathrm{tr} f(P)\right] + (1-\beta)v E \right.$$

$$\left. + v\lambda h(\bar{P}) + (1-v\lambda)h(f(P)) \right\},$$

(3.10)

where ρ is the optimal average cost per stage, and $h(.)$ is the differential cost or relative value function [25]. For the infinite horizon problem (3.9), existence of optimal stationary policies can be ensured via the following result:

Lemma 3.1 *Let $\lambda > 1 - \frac{1}{\max_i |\sigma_i(A)|^2}$, where $\max_i |\sigma_i(A)|$ denotes the spectral radius of A. Then there exists a constant ρ and a function $h(.)$ that satisfies the Bellman equation (3.10).*

Proof See Appendix.

As a consequence of Lemma 3.1, Problem (3.9) can be solved using methods such as the relative value iteration algorithm [25]. In computations, since the state space is (countably) infinite, one commonly first truncates the state space to

$$\mathbb{S}^N \triangleq \{\bar{P}, f(\bar{P}), f^2(\bar{P}), \ldots, f^{N-1}(\bar{P})\}, \text{ for some } N \in \mathbb{N},$$

and then uses the relative value iteration algorithm to solve the resulting finite state space MDP problem, as follows: For a given N, define for $t = 0, 1, 2, \ldots$ the value functions $V_t(.) : \mathbb{S}^N \to \mathbb{R}$ by:

$$V_{t+1}(P) = \min_{\nu \in \{0,1\}} \left\{ \beta[\nu\lambda\mathrm{tr}\bar{P} + (1 - \nu\lambda)\mathrm{tr}f(P)] + (1 - \beta)\nu E \right.$$
$$\left. + \nu\lambda V_t(\bar{P}) + (1 - \nu\lambda)V_t(f(P)) \right\}.$$

Let $P_f \in \mathbb{S}^N$ be fixed. The relative value iteration algorithm is given by:

$$\begin{aligned}
h_{t+1}(P) &\triangleq V_{t+1}(P) - V_{t+1}(P_f) \\
&= \min_{\nu \in \{0,1\}} \left\{ \beta[\nu\lambda\mathrm{tr}\bar{P} + (1 - \nu\lambda)\mathrm{tr}f(P)] + (1 - \beta)\nu E \right. \\
&\quad \left. + \nu\lambda h_t(\bar{P}) + (1 - \nu\lambda)h_t(f(P)) \right\} \\
&\quad - \min_{\nu_f \in \{0,1\}} \left\{ \beta[\nu_f\lambda\mathrm{tr}\bar{P} + (1 - \nu_f\lambda)\mathrm{tr}f(P)] + (1 - \beta)\nu_f E \right. \\
&\quad \left. + \nu_f\lambda h_t(\bar{P}) + (1 - \nu_f\lambda)h_t(f(P)) \right\}.
\end{aligned} \tag{3.11}$$

As $t \to \infty$, we have $h_t(P) \to h(P), \forall P \in \mathbb{S}^N$, with $h(.)$ satisfying the Bellman equation (3.10). In practice, the algorithm (3.11) terminates once the differences $h_{t+1}(P) - h_t(P)$ become smaller than a desired level of accuracy. One then compares the solutions obtained as N increases to determine an appropriate value of N for truncation of the state space $\mathbb{S}$, see Chap. 8 of [26] for further details.

3.1.3 Structural Properties of Optimal Transmission Scheduling

In this subsection, we will derive some structural results on the optimal solutions to the finite horizon problem (3.6) and the infinite horizon problem (3.9). In particular, we will prove that a threshold policy is optimal and derive simple analytical expressions for the expected energy usage and expected error covariance. Knowing that the optimal policy is a threshold policy allows for ease of real-time implementation, and can also allow for more computationally efficient algorithms in the numerical solution of problems (3.6) and (3.9), see [27, 28].

Notation: For symmetric matrices X and Y, we say that $X \leq Y$ if $Y - X$ is positive semi-definite, and $X < Y$ if $Y - X$ is positive definite. Let **S** denote the set of all positive semi-definite matrices.

Definition 3.1 A function $F(.) : \mathbf{S} \to \mathbb{R}$ is *increasing* if

$$X \leq Y \Rightarrow F(X) \leq F(Y). \tag{3.12}$$

Lemma 3.2 *The function* $\mathrm{tr}\, f(X) = \mathrm{tr}(AXA^T + Q)$ *is an increasing function of* X.

Proof This is easily seen from the definition.   □

Recall the set $\mathbb{S}$ defined in (3.5).

Lemma 3.3 *There is a total ordering on the elements of* $\mathbb{S}$ *given by*

$$\bar{P} \leq f(\bar{P}) \leq f^2(\bar{P}) \leq \cdots .$$

Proof This result is proved in [29], see also [30].   □

Theorem 3.1 (i) *The optimal solution to the finite horizon problem (3.6) is of the form:*

$$v_k^* = \begin{cases} 0 \,, & \text{if } P_{k-1|k-1} < P_k^{\mathrm{th}} \\ 1 \,, & \text{if } P_{k-1|k-1} \geq P_k^{\mathrm{th}} \end{cases}$$

for some thresholds $P_k^{\mathrm{th}}, k = 1, \ldots, K$, *where the thresholds may be infinite (meaning that* $v_k^* = 0, \forall P_{k-1|k-1} \in \mathbb{S}$) *when A is stable.*
(ii) *The optimal solution to the infinite horizon problem (3.9) is of the form:*

$$v_k^* = \begin{cases} 0 \,, & \text{if } P_{k-1|k-1} < P^{\mathrm{th}} \\ 1 \,, & \text{if } P_{k-1|k-1} \geq P^{\mathrm{th}} \end{cases} \tag{3.13}$$

for some constant threshold P^{th}, *where the threshold may be infinite when A is stable.*

Proof (i) Since v_k takes on either the values 0 or 1, $J_k(P)$ can be rewritten as

$$J_k(P) = \min \Big\{ \beta \mathrm{tr}(f(P)) + J_{k+1}(f(P)), \ \beta[\lambda \mathrm{tr}(\bar{P}) + (1 - \lambda)\mathrm{tr}(f(P))] + (1 - \beta)E$$
$$+ \lambda J_{k+1}(\bar{P}) + (1 - \lambda)J_{k+1}(f(P)) \Big\}$$

with the two terms in the minimization corresponding to the cases $v_k = 0$ and $v_k = 1$. Let

$$\phi_k(P) \triangleq \beta[\lambda \mathrm{tr}(f(P)) - \lambda \mathrm{tr}(\bar{P})] - (1 - \beta)E + \lambda J_{k+1}(f(P)) - \lambda J_{k+1}(\bar{P}),$$
$$\tag{3.14}$$

which denotes the difference between the two terms, considered as a function of P.
Note that if $\phi_k(P) < 0$ then the sensor will not transmit, while if $\phi_k(P) > 0$ then
the sensor will transmit.

Theorem 3.1 (i) will be proved if we can show that the function $\phi_k(P)$ is an increasing function of P, for $k = 0, \ldots, N - 1$. By Lemma 3.2, $\mathrm{tr}\, f(P)$ is an increasing
function of P. Furthermore, a simple induction argument shows that $J_{k+1}(f(P))$ is
increasing in P. Hence the function $\phi_k(P)$ defined by (3.14) is increasing in P.
(ii) Recalling the relative value iteration algorithm (3.11), one can show using similar arguments as in the proof of (i) that the properties in Theorem 3.1 (i) also hold
when $J_{k+1}(.)$ is replaced with $h_l(.)$. Since $h_l(P) \to h(P)$ as $l \to \infty$, the result then
follows. $\square$

Remark 3.1 In Theorem 3.1, we could have P_k^{th} or P^{th} equal to $\bar{P}$, in which case
$v_k^* = 1, \forall P_{k-1|k-1} \in \mathbb{S}$.

Remark 3.2 Theorem 3.1 was originally proved in [31] using the theory of submodular functions. Under a related set-up that minimizes an expected error covariance
measure subject to a constraint on the communication rate, the optimality of threshold policies over an infinite horizon was also proved using different techniques in
[32].

By Theorem 3.1, the optimal policy is a threshold policy on the error covariance.
This also allows us to derive simple analytical expressions for the expected energy
usage and expected error covariance for the single sensor case over an infinite horizon.
A similar analysis can be carried out for the finite horizon situation but the expressions
will be rather complicated due to the thresholds P_k^{th} in Theorem 3.1 being time-
varying in general.

Let $t \in \mathbb{N}$ be such that $f^t(\bar{P}) = P^{\mathrm{th}} \in \mathbb{S}$, see (3.13). Note that t will depend on the
value of β chosen in problem (3.9). Then the evolution of the error covariance at the
remote estimator can be modelled as the (countably infinite) Markov chain shown
in Fig. 3.2, where state i of the Markov chain corresponds to the value $f^i(\bar{P})$, $i =
0, 1, 2, \ldots$. The transition probability matrix **P** for this Markov chain can be written
as:

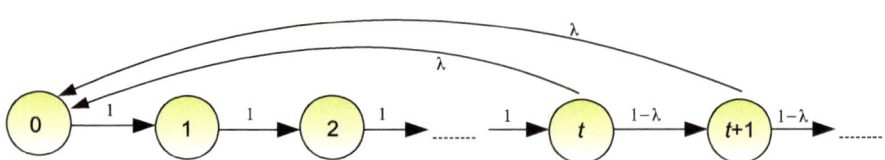

Fig. 3.2 Markov chain for threshold policy

$$\mathbf{P} = \begin{bmatrix} 0 & 1 & 0 & \ldots\ldots & & \ldots \\ 0 & 0 & 1 & 0 & \ldots & & \ldots \\ \vdots & & \ddots & \ddots & & & \\ 0 & \ldots\ldots & 0 & 1 & 0 & & \ldots \\ \lambda & 0 & \ldots & & 0 & 1-\lambda & 0 & \ldots\ldots \\ \lambda & 0 & \ldots & & & 0 & 1-\lambda & 0 & \ldots \\ \vdots & \vdots & & & & & & \ddots \end{bmatrix}.$$

For $\lambda \in (0, 1)$, one can easily verify that this Markov chain is irreducible, aperiodic and with all states being positive recurrent. Then the stationary distribution

$$\pi = \begin{bmatrix} \pi_0 & \pi_1 & \pi_2 & \ldots & \pi_t & \pi_{t+1} & \pi_{t+2} & \ldots \end{bmatrix},$$

where π_j is the stationary probability of the Markov chain being in state j, exists and can be computed using the relation $\pi = \pi\mathbf{P}$. We find after some calculations that $\pi_j = \pi_0$, $j = 1, \ldots, t$, and $\pi_j = (1 - \lambda)^{j-t}\pi_0$, $j = t + 1, t + 2, \ldots$, and thereby

$$\pi_0 = \frac{1}{t + 1/\lambda} = \frac{\lambda}{\lambda t + 1}.$$

Hence

$$\pi_j = \begin{cases} \frac{\lambda}{\lambda t+1}, & \text{if } j = 0, \ldots, t \\ \frac{(1-\lambda)^{j-t}\lambda}{\lambda t+1}, & \text{if } j = t + 1, t + 2, \ldots. \end{cases}$$

We can now derive analytical expressions for the expected energy usage and expected error covariance. For the expected energy usage, since the sensor transmits only when the Markov chain is in states $t, t + 1, \ldots$, an energy amount of E is used in reaching the states corresponding to $\bar{P}$, $f^{t+1}(\bar{P})$, $f^{t+2}(\bar{P})$, $\ldots$. Hence

$$\begin{align} \mathbb{E}[\text{energy}] &= E[\pi_0 + \pi_{t+1} + \pi_{t+2} + \cdots] \\ &= E\pi_0[1 + 1 - \lambda + (1 - \lambda)^2 + \cdots] \\ &= \frac{E\pi_0}{\lambda} = \frac{E}{\lambda t + 1}. \end{align} \tag{3.15}$$

For the expected error covariance, we have

$$\mathbb{E}[\text{tr} P_{k|k}] = \pi_0 \text{tr}(\bar{P}) + \pi_1 \text{tr}(f(\bar{P})) + \pi_2 \text{tr}(f^2(\bar{P})) + \cdots \tag{3.16}$$

which can be computed numerically. Under the assumption that $\lambda > 1 - \frac{1}{\max_i |\sigma_i(A)|^2}$, $\mathbb{E}[\text{tr} P_{k|k}]$ will be finite, by a similar argument as that used in the proof of Lemma 3.1.

3.1.4 Transmission of Measurements

In this subsection, we will study the situation where, as in Chap. 2, sensor measurements instead of local state estimates are transmitted to the remote estimator. In particular, we wish to derive structural results on the optimal transmission schedule. Our descriptions of the model and optimization problem below will be kept brief, in order to proceed quickly to the structural results.

System Model

The process and measurements follow the same model as in (3.1) and (3.2). Let $v_k \in \{0, 1\}$ be decision variables such that $v_k = 1$ if the measurement y_k (rather than the local state estimate) is to be transmitted to the remote estimator at time k, and $v_k = 0$ if there is no transmission. As before (see Fig. 3.1), the transmit decisions v_k are to be decided at the remote estimator and assumed to only depend on the error covariance at the remote estimator.

At the remote estimator, if no sensors are scheduled to transmit, then the state estimates and error covariances are updated by:

$$
\begin{aligned}
\hat{x}_{k+1|k} &= A\hat{x}_{k|k} \\
\hat{x}_{k|k} &= \hat{x}_{k|k-1} \\
P_{k+1|k} &= A P_{k|k} A^T + Q \\
P_{k|k} &= P_{k|k-1}.
\end{aligned}
\tag{3.17}
$$

If the sensor has been scheduled by the remote estimator to transmit at time k, then the state estimates and error covariances at the remote estimator are now updated as follows:

$$
\begin{aligned}
\hat{x}_{k+1|k} &= A\hat{x}_{k|k} \\
\hat{x}_{k|k} &= \hat{x}_{k|k-1} + \gamma_k K_k (y_k - C\hat{x}_{k|k-1}) \\
P_{k+1|k} &= A P_{k|k} A^T + Q \\
P_{k|k} &= P_{k|k-1} - \gamma_k K_k C P_{k|k-1},
\end{aligned}
\tag{3.18}
$$

where $K_k \triangleq P_{k|k-1}C^T (C P_{k|k-1} C^T + R)^{-1}$. We can thus write:

$$
P_{k+1|k} =
\begin{cases}
f(P_{k|k-1}), & \text{if } v_k \gamma_k = 0 \\
g(P_{k|k-1}), & \text{if } v_k \gamma_k = 1,
\end{cases}
\tag{3.19}
$$

where $f(X) = A X A^T + Q$ as before, and

$$
g(X) \triangleq A X A^T - A X C^T (C X C^T + R)^{-1} C X A^T + Q.
\tag{3.20}
$$

In (3.19), the recursions are given in terms of $P_{k+1|k}$ rather than $P_{k|k}$, since the resulting expressions are slightly more convenient to work with.

Optimization of Transmission Scheduling

We consider transmission policies $v_k(P_{k|k-1})$ that depend only on $P_{k|k-1}$. The finite horizon optimization problem is:

$$\min_{\{v_k\}} \sum_{k=1}^{K} \mathbb{E}\left[\mathbb{E}\left[\beta \mathrm{tr} P_{k+1|k} + (1-\beta)v_k E \,\middle|\, P_{k|k-1}, v_k\right]\right] \tag{3.21}$$

where we can compute

$$\mathbb{E}[\mathrm{tr} P_{k+1|k}|P_{k|k-1}, v_k] = v_k \lambda \mathrm{tr} g(P_{k|k-1}) + (1 - v_k \lambda)\,\mathrm{tr} f(P_{k|k-1}),$$

with $f(.)$ defined in (3.4) and $g(.)$ defined in (3.20). Let the functions $J_k(.)$ be given by:

$$J_{K+1}(P) = 0$$

$$J_k(P) = \min_{v \in \{0,1\}} \left\{\beta\left[v\lambda \mathrm{tr} g(P) + (1 - v\lambda)\,\mathrm{tr} f(P)\right] + (1-\beta)vE\right.$$

$$\left. + v\lambda J_{k+1}(g(P)) + (1 - v\lambda)\,J_{k+1}(f(P))\right\}, \quad k = K, K-1, \ldots, 1. \tag{3.22}$$

The infinite horizon problem can be formulated in a similar manner, but will be omitted for brevity.

Structural Properties of Optimal Transmission Scheduling

We will here prove Theorem 3.2, which is the counterpart of Theorem 3.1 for scalar systems when measurements are transmitted, and in particular, establishes the optimality of threshold policies. However, for vector systems we will give a counterexample (Example 3.1) to show that, in general, the optimal policy is not a simple threshold policy. Lemma 3.4 and Theorem 3.2 below will assume scalar systems, thus A, C, Q, R and P are all scalar.

Lemma 3.4 *Let $\mathscr{F}(.)$ be a function formed by composition (in any order) of any of the functions $f(.), g(.),$ and $\mathrm{id}(.),$ where*

$$f(P) \triangleq A^2 P + Q, \quad g(P) \triangleq A^2 P + Q - \frac{A^2 C^2 P^2}{C^2 P + R},$$

and $\mathrm{id}(.)$ is the identity function. Then:
(i) $\mathscr{F}(.)$ is either of the affine form

$$\mathscr{F}(P) = aP + b, \text{ for some } a, b \geq 0, \tag{3.23}$$

or the linear fractional form

$$\mathscr{F}(P) = \frac{aP+b}{cP+d}, \text{ for some } a, b, c, d \geq 0 \text{ with } ad - bc \geq 0. \tag{3.24}$$

(ii) $\mathscr{F}(f(P)) - \mathscr{F}(g(P))$ is an increasing function of P.

Proof (i) We prove this by induction. Firstly, $\mathrm{id}(P) = P$ has the form (3.23), $f(P) = A^2 P + Q$ has the form (3.23), and

$$g(P) = \frac{(A^2 R + C^2 Q)P + RQ}{C^2 P + R}$$

has the form (3.24) since $(A^2 R + C^2 Q)R - RQC^2 = A^2 R^2 \geq 0$.

Now assume that $\mathscr{F}(.)$, which is a composition of the functions $f(.)$, $g(.)$, and $\mathrm{id}(.)$, has the form of either (3.23) or (3.24). Then we will show that $f(\mathscr{F}(P))$ and $g(\mathscr{F}(P))$ also has the form of either (3.23) or (3.24). For notational convenience, let us write $f(P) = \bar{a}P + \bar{b}$ for some $\bar{a}, \bar{b} \geq 0$, and $g(P) = (\bar{a}P + \bar{b})/(\bar{c}P + \bar{d})$ for some $\bar{a}, \bar{b}, \bar{c}, \bar{d} \geq 0$ with $\bar{a}\bar{d} - \bar{b}\bar{c} \geq 0$, which can be achieved as shown at the beginning of the proof.

If $\mathscr{F}(.)$ has the form (3.23), then

$$f(\mathscr{F}(P)) = \bar{a}(aP + b) + \bar{b}$$

is of the form (3.23), and

$$g(\mathscr{F}(P)) = \frac{\bar{a}(aP+b)+\bar{b}}{\bar{c}(aP+b)+\bar{d}} = \frac{\bar{a}aP + \bar{a}b + \bar{b}}{\bar{c}aP + \bar{c}b + \bar{d}}$$

has the form (3.24), since $\bar{a}a(\bar{c}b + \bar{d}) - (\bar{a}b + \bar{b})\bar{c}a = a(\bar{a}\bar{d} - \bar{b}\bar{c}) \geq 0$.

If $\mathscr{F}(.)$ has the form (3.24), then

$$f(\mathscr{F}(P)) = \frac{\bar{a}(aP+b)}{cP+d} + \bar{b} = \frac{(\bar{a}a + \bar{b}c)P + \bar{a}b + \bar{b}d}{cP+d}$$

has the form (3.24), since $(\bar{a}a + \bar{b}c)d - (\bar{a}b + \bar{b}d)c = \bar{a}(ad - bc) \geq 0$. Finally,

$$g(\mathscr{F}(P)) = \frac{\bar{a}\left(\frac{aP+b}{cP+d}\right) + \bar{b}}{\bar{c}\left(\frac{aP+b}{cP+d}\right) + \bar{d}} = \frac{(\bar{a}a + \bar{b}c)P + \bar{a}b + \bar{b}d}{(\bar{c}a + \bar{d}c)P + \bar{c}b + \bar{d}d}$$

has the form (3.24), since $(\bar{a}a + \bar{b}c)(\bar{c}b + \bar{d}d) - (\bar{a}b + \bar{b}d)(\bar{c}a + \bar{d}c) = (ad - bc)(\bar{a}\bar{d} - \bar{b}\bar{c}) \geq 0$.

(ii) By part (i), we know that $\mathscr{F}(.)$ is either of the form (3.23) or (3.24). If $\mathscr{F}(.)$ has the form (3.23), then

$$\mathscr{F}(f(P)) - \mathscr{F}(g(P)) = a(f(P) - g(P))$$

will be an increasing function of P, since $f(P) - g(P) = (A^2C^2P^2)/(C^2P + R)$ can be easily checked to be an increasing function of P.

If $\mathscr{F}(.)$ has the form (3.24), then it can be verified after some algebra that

$$\frac{d}{dP}(\mathscr{F}(f(P)) - \mathscr{F}(g(P))) = \frac{d}{dP}\left(\frac{af(P)+b}{cf(P)+d} - \frac{ag(P)+b}{cg(P)+d}\right)$$

$$= \frac{(ad-bc)A^2C^2P(d+cQ)}{(d + c(A^2P + Q))^2} \frac{C^2P(d+cQ)+2(d+c(A^2P+Q))R}{\left(C^2P(d + cQ) + (d + c(A^2P + Q))R\right)^2} \geq 0$$

since $ad - bc \geq 0$. Hence $\mathscr{F}(f(P)) - \mathscr{F}(g(P))$ is an increasing function of P. □

Theorem 3.2 *The optimal solution to problem (3.21) is of the form:*

$$v_k^* = \begin{cases} 0 \text{ , if } P_{k-1|k-1} < \tilde{P}_k^{\text{th}} \\ 1 \text{ , if } P_{k-1|k-1} \geq \tilde{P}_k^{\text{th}} \end{cases}$$

for some thresholds $\tilde{P}_k^{\text{th}}, k = 1, \ldots, K$.

Proof Define the functions

$$\phi_k(P) \triangleq \beta f(P) + J_{k+1}(f(P)) - \beta[\lambda g(P) + (1-\lambda)f(P)]$$
$$- (1-\beta)E - \lambda J_{k+1}(g(P)) - (1-\lambda)J_{k+1}(f(P))$$

Similar to Theorem 3.1, Theorem 3.2 will be proved if we can show that $\phi_k(P)$ are increasing functions of P. Note that the functions are equivalent to

$$\phi_k(P) = \beta\lambda[f(P) - g(P)] - (1 - \beta)E + \lambda[J_{k+1}(f(P)) - J_{k+1}(g(P))].$$
$$(3.25)$$

As stated in the proof of Lemma 3.4 (ii), $f(P) - g(P)$ can be easily verified to be an increasing function of P. Thus Theorem 3.2 follows if we can show that $J_k(f(P)) - J_k(g(P))$ is an increasing function of P for all k. The proof is by induction. In order to make the induction argument work, we will prove the slightly stronger statement that $J_k(\mathscr{F}(f(P))) - J_k(\mathscr{F}(g(P)))$ is an increasing function of P for all k, where $\mathscr{F}(.)$ is a function formed by composition of any of the functions $f(.), g(.), \text{id}(.)$. The case of $J_{K+1}(\mathscr{F}(f(.))) - J_{K+1}(\mathscr{F}(g(.))) = 0$ is clear. Now assume that for $P' \geq P$,

$$J_{k'}(\mathscr{F}(f(P'))) - J_{k'}(\mathscr{F}(g(P'))) - J_{k'}(\mathscr{F}(f(P))) + J_{k'}(\mathscr{F}(g(P))) \geq 0$$

holds for $k' - K + 1, K, \ldots, k + 1$. We have

$$J_k(\mathscr{F}(f(P'))) - J_k(\mathscr{F}(g(P'))) - J_k(\mathscr{F}(f(P))) + J_k(\mathscr{F}(g(P)))$$

$$\geq \min_{v \in \{0,1\}} \Big\{ \beta \Big[v\lambda g(\mathscr{F}(f(P'))) + (1 - v\lambda) f(\mathscr{F}(f(P'))) \Big]$$

$$+ v\lambda J_{k+1}(g(\mathscr{F}(f(P')))) + (1 - v\lambda) J_{k+1}(f(\mathscr{F}(f(P'))))$$

$$- \beta \Big[v\lambda g(\mathscr{F}(g(P'))) + (1 - v\lambda) f(\mathscr{F}(g(P'))) \Big]$$

$$- v\lambda J_{k+1}(g(\mathscr{F}(g(P')))) - (1 - v\lambda) J_{k+1}(f(\mathscr{F}(g(P'))))$$

$$- \beta \Big[v\lambda g(\mathscr{F}(f(P))) + (1 - v\lambda) f(\mathscr{F}(f(P))) \Big]$$

$$- v\lambda J_{k+1}(g(\mathscr{F}(f(P)))) - (1 - v\lambda) J_{k+1}(f(\mathscr{F}(f(P))))$$

$$+ \beta \Big[v\lambda g(\mathscr{F}(g(P))) + (1 - v\lambda) f(\mathscr{F}(g(P))) \Big]$$

$$+ v\lambda J_{k+1}(g(\mathscr{F}(g(P)))) + (1 - v\lambda) J_{k+1}(f(\mathscr{F}(g(P)))) \Big\}. \tag{3.26}$$

In the minimization of (3.26) above, if the optimal $v^* = 0$, then

$$J_k(\mathscr{F}(f(P'))) - J_k(\mathscr{F}(g(P'))) - J_k(\mathscr{F}(f(P))) + J_k(\mathscr{F}(g(P)))$$

$$\geq \beta \big[f(\mathscr{F}(f(P'))) - f(\mathscr{F}(g(P'))) - f(\mathscr{F}(f(P))) + f(\mathscr{F}(g(P))) \big]$$

$$+ J_{k+1}(f(\mathscr{F}(f(P')))) - J_{k+1}(f(\mathscr{F}(g(P'))))$$

$$- J_{k+1}(f(\mathscr{F}(f(P)))) + J_{k+1}(f(\mathscr{F}(g(P)))) \geq 0$$

where the last inequality holds by Lemma 3.4 (ii), the induction hypothesis, and the fact that $f \circ \mathscr{F}(.)$ is a composition of functions of the form $f(.)$, $g(.)$, id(.). If instead the optimal $v^* = 1$, then by a similar argument

$$J_k(\mathscr{F}(f(P'))) - J_k(\mathscr{F}(g(P'))) - J_k(\mathscr{F}(f(P))) + J_k(\mathscr{F}(g(P)))$$

$$\geq \beta\lambda_l \big[g(\mathscr{F}(f(P'))) - g(\mathscr{F}(g(P'))) - g(\mathscr{F}(f(P))) + g(\mathscr{F}(g(P))) \big]$$

$$+ \beta(1 - \lambda_l) \big[f(\mathscr{F}(f(P'))) - f(\mathscr{F}(g(P'))) - f(\mathscr{F}(f(P))) + f(\mathscr{F}(g(P))) \big]$$

$$+ \lambda_l \big[J_{k+1}(g(\mathscr{F}(f(P')))) - J_{k+1}(g(\mathscr{F}(g(P'))))$$

$$- J_{k+1}(g(\mathscr{F}(f(P)))) + J_{k+1}(g(\mathscr{F}(g_m(P)))) \big]$$

$$+ (1 - \lambda_l) \big[J_{k+1}(f(\mathscr{F}(f(P')))) - J_{k+1}(f(\mathscr{F}(g(P'))))$$

$$- J_{k+1}(f(\mathscr{F}(f(P)))) + J_{k+1}(f(\mathscr{F}(g(P)))) \big] \geq 0, \tag{3.27}$$

from where the result follows. $\quad \square$

Theorem 3.2 is the counterpart of Theorem 3.1 for estimation schemes where the system is scalar and measurements are transmitted. It provides a theoretical justification for the variance-based triggering strategy proposed in [7]. For vector systems, Theorem 3.2, in general, does not hold, as the following counterexample shows.

Example 3.1 Consider the case $K = 1$, corresponding to the case of a single transmission. Suppose we have a system with parameters

$$A = \begin{bmatrix} 1.1 & 0.2 \\ 0.2 & 0.8 \end{bmatrix}, \quad C = \begin{bmatrix} 1 & -0.9 \end{bmatrix},$$

$Q = I$, $R = 1$, $\beta = 0.5$, $\lambda = 0.7$, $E = 0.85$. Let

$$P = \bar{P} = \begin{bmatrix} 7.8328 & 7.3915 \\ 7.3915 & 7.7127 \end{bmatrix}, \quad P' = \begin{bmatrix} 7.85 & 7.40 \\ 7.40 & 7.80 \end{bmatrix}.$$

Then one can easily verify that $P' > P$, but that $\phi(P') < 0$ and $\phi(P) > 0$, so that one transmits for the smaller value P, but not for the larger value P'. $\square$

For vector systems with scalar measurements, a threshold policy was considered in [7], where a sensor would transmit if CPC^T exceeded a threshold. Since $P' > P$ implies $CP'C^T > CPC^T$, the above example also shows that such a threshold policy is in general not optimal when measurements are transmitted, under our problem formulation of minimizing a linear combination of the expected error covariance and expected energy usage.

3.1.5 Numerical Studies

We consider a system of the form (3.1) and (3.2) with parameters

$$A = \begin{bmatrix} 1.1 & 0.2 \\ 0.2 & 0.8 \end{bmatrix}, \quad C = \begin{bmatrix} 1 & 1 \end{bmatrix}, \quad Q = I, \quad R = 1,$$

in which case $\bar{P}$ is easily computed as

$$\bar{P} = \begin{bmatrix} 1.3762 & -0.9014 \\ -0.9014 & 1.1867 \end{bmatrix}.$$

The packet reception probability is chosen to be $\lambda = 0.8$, and the transmission energy is taken as $E = 1$.

We first consider the finite horizon problem (3.6), with $K = 5$ and $\beta = 0.05$. Figures 3.3 and 3.4 plot respectively the optimal v_1^* and v_2^* (i.e. $k = 1$ and $k = 2$) for different values of $f^n(\bar{P})$, which we recall represent the different values that the error covariance can take. In agreement with Theorem 3.1, we observe a threshold behaviour in the optimal v_k^*. In this example, we have $P_1^{\text{th}} = f^3(\bar{P})$ and $P_2^{\text{th}} = f^2(\bar{P})$; the thresholds are in general different for different values of k.

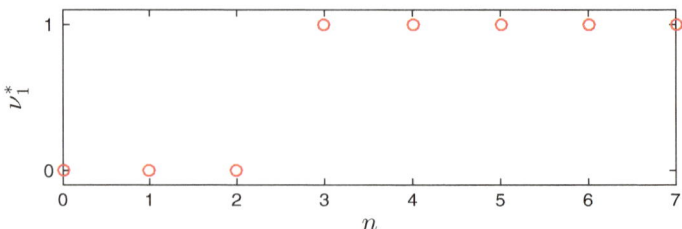

Fig. 3.3 Finite horizon, $K = 5$. v_1^* for different values of $f^n(\bar{P})$

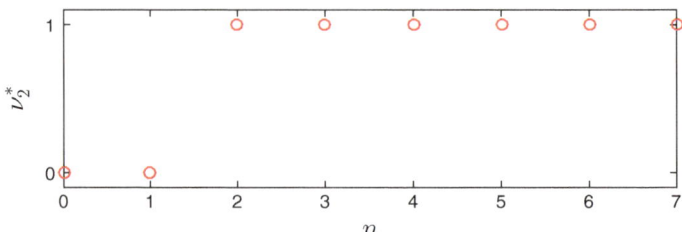

Fig. 3.4 Finite horizon, $K = 5$. v_2^* for different values of $f^n(\bar{P})$

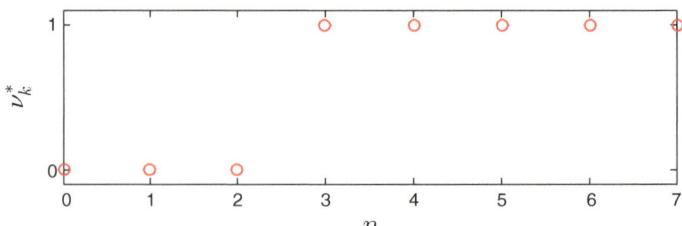

Fig. 3.5 Infinite horizon. v_k^* for different values of $f^n(\bar{P})$

We next consider the infinite horizon problem (3.9), with $\beta = 0.05$. Figure 3.5 plots the optimal v_k^* for different values of $f^n(\bar{P})$, where we again see a threshold behaviour, with $P^{\text{th}} = f^3(\bar{P})$. In Fig. 3.6 we plot the values of the thresholds for different values of β. As β increases, the relative importance of minimizing the error covariance (vs the energy usage) is increased, thus one should transmit more often, leading to decreasing values of the thresholds.

Finally, in Fig. 3.7 we plot the trace of the expected error covariance versus the expected energy, obtained by solving the infinite horizon problem for different values of β, with the values computed using the expressions (3.15) and (3.16). We observe that a smaller expected error covariance can be obtained for higher expected energy usage. Note that the plot is discrete as $t \in \mathbb{N}$ in (3.15) and (3.16), see also Fig. 3.6.

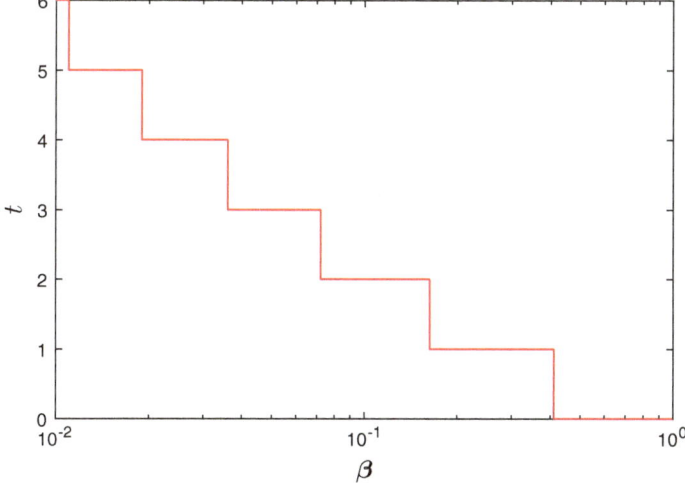

Fig. 3.6 Infinite horizon. Threshold P^{th} versus β, with $f^t(\bar{P}) = P^{\text{th}}$

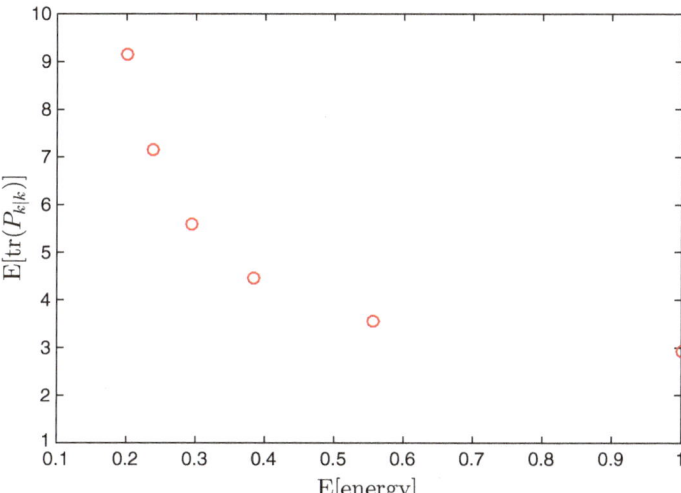

Fig. 3.7 Infinite horizon. Expected error covariance versus expected energy

3.2 Transmission Scheduling with Energy Harvesting

3.2.1 System Model

A diagram of the system model for this section is shown in Fig. 3.8. The discrete-time process and sensor measurements are as given in (3.1) and (3.2). As before,

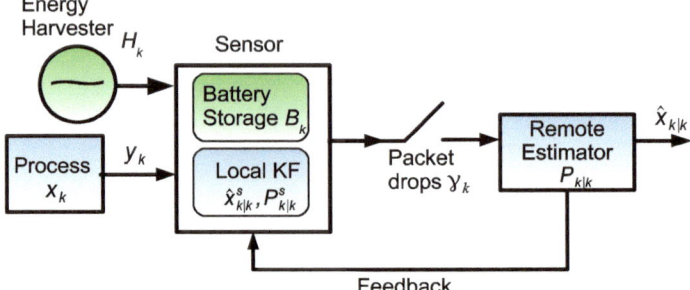

Fig. 3.8 Transmission scheduling with an energy harvesting sensor

let $v_k \in \{0, 1\}$ be decision variables such that $v_k = 1$ if and only if $\hat{x}^s_{k|k}$ is to be transmitted to the remote estimator at time k. Let B_k denote the battery level of the sensor at time k, with $B_{\max}$ the maximum capacity of the battery. There is an energy usage of E for each scheduled transmission. Here transmission can only occur if there is sufficient battery energy, i.e. $v_k = 1$ is possible only when the battery level $B_k \geq E$. The sensor is equipped with energy harvesting capabilities, with the energy harvested between the discrete-time instants $k - 1$ and k denoted by H_k. Similar to [33], the evolution of the battery level is modelled as

$$B_{k+1} = \min\{B_k - v_k E + H_{k+1}, B_{\max}\} = \mathfrak{g}(B_k - v_k E + H_{k+1}) \qquad (3.28)$$

with $v_k = 0$ if $B_k < E$, and where the function $\mathfrak{g}(.)$ is defined by

$$\mathfrak{g}(x) \triangleq \min\{x, B_{\max}\}. \qquad (3.29)$$

The harvested energy process $\{H_k\}$ is random and here assumed to be a Markov process, with state space $\mathbb{H}$. Also denote $\mathbb{B} \triangleq [0, B_{\max}]$. The decision variables v_k are determined at the sensor, and will be assumed to depend on $P_{k-1|k-1}$, H_k and B_k. We assume that γ_{k-1} is fed back to the sensor before the transmission decision at time k. Thus, the remote estimator error covariance $P_{k-1|k-1}$ can be reconstructed at the sensor with this acknowledgement mechanism.[4]

The information set available to the remote estimator is again

$$\mathbb{I}_k \triangleq \{v_0, \ldots, v_k, v_0\gamma_0, \ldots, v_k\gamma_k, v_0\gamma_0\hat{x}^s_{0|0}, \ldots, v_k\gamma_k\hat{x}^s_{k|k}\}.$$

Given that the decision variables v_k depend on $P_{k-1|k-1}$, H_k and B_k, but not on the current state x_k or measurement y_k, the optimal remote estimator will also be of the form (3.3).

---

[4]The case of imperfect feedback acknowledgements can also be considered, using similar ideas as in [34].

Remark 3.3 Note that even if there are no packet drops, i.e. $\gamma_k = 1$, $\forall k$, $\{P_{k|k}\}$ will in general still be a random process, due to the random nature of the harvested energy which place constraints on whether a transmission can occur. In contrast, when there are no packet drops and no energy harvesting constraints, $\{P_{k|k}\}$ will be deterministic under threshold policies on the estimation error covariance [7].

3.2.2 Optimization of Transmission Scheduling

As mentioned in Sect. 3.2.1, we will consider transmission policies where ν_k depends only on $P_{k-1|k-1}$, H_k and B_k. We will consider the following optimization problem of finite horizon K:

$$\min_{\substack{\nu_k \in \{0,1\} \\ \nu_k E \leq B_k}} \sum_{k=1}^{K} \mathbb{E}[\operatorname{tr} P_{k|k}] = \min_{\substack{\nu_k \in \{0,1\} \\ \nu_k E \leq B_k}} \sum_{k=1}^{K} \mathbb{E}\big[\mathbb{E}[\operatorname{tr} P_{k|k} | P_{k-1|k-1}, \nu_k, H_k, B_k]\big]. \tag{3.30}$$

We note that

$$\mathbb{E}[\operatorname{tr} P_{k|k} | P_{k-1|k-1}, \nu_k, H_k, B_k] = \nu_k \lambda \operatorname{tr}(\bar{P}) + (1 - \nu_k \lambda) \operatorname{tr} f(P_{k-1|k-1}).$$

Let the functions $J_k(\cdot, \cdot, \cdot) : \mathbb{S} \times \mathbb{H} \times \mathbb{B} \to \mathbb{R}$ be defined recursively as:

$$J_{K+1}(P, H, B) = 0$$

$$\begin{aligned}
J_k(P, H, B) = \min_{\substack{\nu \in \{0,1\} \\ \nu E \leq B}} \Big\{ &\nu \lambda \operatorname{tr}(\bar{P}) + (1 - \nu \lambda) \operatorname{tr} f(P) \\
&+ \nu \lambda \mathbb{E}\Big[J_{K+1}(\bar{P}, \tilde{H}, \mathfrak{g}(B - \nu E + \tilde{H})) | H \Big] \\
&+ (1 - \nu \lambda) \mathbb{E}\Big[J_{K+1}(f(P), \tilde{H}, \mathfrak{g}(B - \nu E + \tilde{H})) | H \Big] \Big\}, \quad k = K, \dots, 1,
\end{aligned} \tag{3.31}$$

where the conditional expectations are with respect to $\tilde{H}$ given H, and $\mathfrak{g}(.)$ is defined in (3.29). Problem (3.30) can be solved using the dynamic programming algorithm, by computing $J_k(P_{k-1|k-1}, H_k, B_k)$ for $k = K, K-1, \dots, 1$. Note that if the range of H_k is continuous, then, in practice, H_k and B_k will need to be discretized in order for problem (3.30) to be solved numerically.

3.2.3 Structural Properties of Optimal Transmission Scheduling

In this subsection, we will show that for a given B_k and H_k, the optimal policy is a threshold policy with respect to the error covariance $P_{k-1|k-1}$, i.e. it is optimal to transmit if and only if $P_{k-1|k-1}$ exceeds a certain threshold (dependent on k, B_k and H_k). Similarly, for fixed $P_{k-1|k-1}$ and H_k, the optimal policy is a threshold policy with respect to the battery level B_k.

Recall from Definition 3.1 that a function $F(.) : \mathbb{S} \to \mathbb{R}$ is increasing if

$$X \leq Y \Rightarrow F(X) \leq F(Y). \tag{3.32}$$

Lemma 3.5 *For any $n \in \mathbb{N}$, $\operatorname{tr} f^n(P)$ is an increasing function of P.*

Proof We have

$$\operatorname{tr} f^n(P) = \operatorname{tr}\left(A^n P (A^n)^T + \sum_{m=0}^{n-1} A^m Q (A^m)^T \right),$$

which is increasing with P. □

Lemma 3.6 *For $d \geq 0$, the function $\mathfrak{g}(.)$ defined in (3.29) satisfies*

$$0 \leq \mathfrak{g}(x) - \mathfrak{g}(x - d) \leq d.$$

Proof The inequality $\mathfrak{g}(x) - \mathfrak{g}(x - d) \geq 0$ is obvious. For the other inequality, note that if $x \leq B_{\max}$, then $\mathfrak{g}(x) - \mathfrak{g}(x - d) = x - (x - d) = d$. If $x > B_{\max}$ and $x - d > B_{\max}$, then $\mathfrak{g}(x) - \mathfrak{g}(x - d) = B_{\max} - B_{\max} = 0$. If $x > B_{\max}$ (which implies $x - d > B_{\max} - d$) and $x - d \leq B_{\max}$, then it holds that $\mathfrak{g}(x) - \mathfrak{g}(x - d) = B_{\max} - (x - d) < B_{\max} - (B_{\max} - d) = d$. □

Theorem 3.3 (i) *For fixed B_k and H_k, the optimal v_k^* is a threshold policy on $P_{k-1|k-1}$ of the form:*

$$v_k^*(P_{k-1|k-1}, B_k, H_k) = \begin{cases} 0 \text{ , if } P_{k-1|k-1} \leq P_k^* \\ 1 \text{ , otherwise,} \end{cases}$$

where the threshold P_k^* depends on k, B_k and H_k.
(ii) *For fixed $P_{k-1|k-1}$ and H_k, the optimal v_k^* is a threshold policy on B_k of the form:*

$$v_k^*(P_{k-1|k-1}, B_k, H_k) = \begin{cases} 0 \text{ , if } B_k \leq B_k^* \\ 1 \text{ , otherwise,} \end{cases}$$

where the threshold B_k^* depends on k, $P_{k-1|k-1}$ and H_k.

Proof See Appendix. □

By making use of both parts (i) and (ii) of Theorem 3.3, we see that for a given k and H_k, the region of possible values of $(P_{k-1|k-1}, B_k)$ can be divided into a 'transmit' and 'don't transmit' region, separated by a staircase-like threshold, see Fig. 3.9. Knowing that the optimal policies are of threshold-type simplifies real-time implementation. In addition, specialized algorithms can be derived which can provide computational

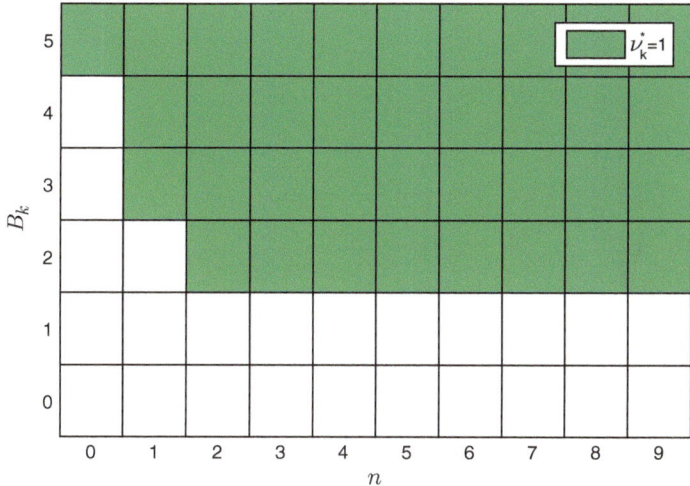

Fig. 3.9 v_k^* for different values of $P_{k-1|k-1} = f^n(\bar{P})$ and B_k, with $k = 5$ and $H_k = 1$

savings when solving problem (3.30) numerically. For example, for a given (k, H_k), suppose we want to compute $v_k^*(P_{k-1|k-1}, B_k, H_k)$ for all possible $(P_{k-1|k-1}, B_k)$. Without structural information, one would need to compare the values $v_k = 0$ and $v_k = 1$ at each $(P_{k-1|k-1}, B_k)$, which results in around $(k + 1) \times |\mathbb{B}|$ comparisons, where $|\mathbb{B}|$ is the cardinality of $\mathbb{B}$. However, by using structural information, one way to determine the staircase-like threshold is as follows. For the smallest possible value of $P_{k-1|k-1}$, which is $\bar{P}$, search (in decreasing order) starting from the largest value of B_k (i.e. $B_{\max}$) to find the threshold $B_k^*(\bar{P}, H_k)$. Then for the next smallest value of $P_{k-1|k-1}$, which is $f(\bar{P})$, the threshold $B_k^*(f(\bar{P}), H_k)$ satisfies $B_k^*(f(\bar{P}), H_k) \leq B_k^*(\bar{P}, H_k)$, so we can now search (in decreasing order) for $B_k^*(f(\bar{P}), H_k)$ starting from $B_k^*(\bar{P}, H_k)$ rather than from $B_{\max}$. Continuing this procedure until we cover all possible values of $P_{k-1|k-1}$, it is not too difficult to see that the number of comparisons required (for each (k, H_k)) to determine the staircase-like threshold is upper bounded by $2(k + 1 + |\mathbb{B}|)$. This could be significantly smaller than $(k + 1) \times |\mathbb{B}|$, the number of comparisons needed if no structural information is assumed.

3.2.4 Numerical Studies

We consider an example with parameters

$$A = \begin{bmatrix} 1.2 & 0.2 \\ 0.2 & 0.7 \end{bmatrix}, \quad C = \begin{bmatrix} 1 & 1 \end{bmatrix}, \quad Q = I, \quad R = 1,$$

in which case

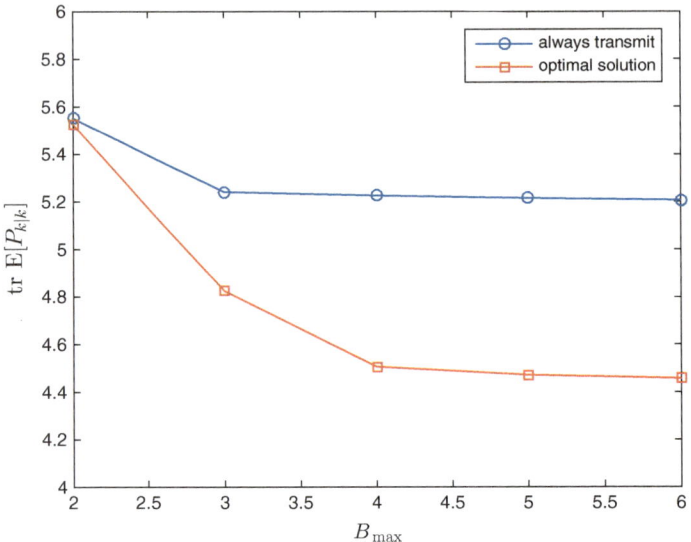

Fig. 3.10 Expected error covariance versus maximum battery capacity

$$\bar{P} = \begin{bmatrix} 1.3634 & -0.8347 \\ -0.8347 & 1.0809 \end{bmatrix}.$$

The packet reception probability is chosen to be $\lambda = 0.7$. The transmission energy is set to $E = 2$. The harvested energy process $\{H_k\}$ is chosen to be a Markov chain with state space $\{0, 1, 2\}$ and transition probability matrix

$$\begin{bmatrix} p_{00} & p_{01} & p_{02} \\ p_{10} & p_{11} & p_{12} \\ p_{20} & p_{21} & p_{22} \end{bmatrix} = \begin{bmatrix} 0.2 & 0.3 & 0.5 \\ 0.3 & 0.4 & 0.3 \\ 0.1 & 0.2 & 0.7 \end{bmatrix},$$

with the initial distribution of (H_1) having the stationary distribution. The maximum battery capacity $B_{max} = 6$. We use the finite horizon $K = 10$. Figure 3.9 plots v_k^* for different values of $P_{k-1|k-1} = f^n(\bar{P})$ and B_k, for fixed $k = 5$ and $H_k = 1$. We observe threshold-like behaviour in agreement with Theorem 3.3.

Next, we study the effect of varying the maximum battery capacity B_{max}. Figure 3.10 plots the trace of the expected error covariance $\mathrm{tr}\mathbb{E}[P_{k|k}]$ versus B_{max}, with $\mathrm{tr}\mathbb{E}[P_{k|k}]$ obtained by averaging over 1,00,000 Monte Carlo runs, with each run having the initial values $P_{0|0} = \bar{P}$ and $B_1 = E$. We compare the performance with a simple suboptimal greedy policy which always transmits provided it has enough energy. We see that performance generally improves as B_{max} increases, though for larger B_{max} further performance gains are small. We also see that the optimal policy significantly outperforms the greedy policy, while the average energy usage (or number of transmissions) of the optimal scheme is no larger than that of the greedy policy.

3.3 Conclusion

This chapter has studied event-based remote estimation problems, with sensor transmissions over a packet dropping channel. By considering an optimization problem for transmission scheduling that minimizes a linear combination of the expected error covariance at the remote estimator and the expected energy across the sensors, we have derived structural properties in the form of the optimal solution, when either local state estimates or sensor measurements are transmitted. In particular, our results show that a threshold policy is optimal.

Later, we considered the case where the sensor is equipped with energy harvesting capabilities. We derived structural results on the optimal transmission scheduling in order to minimize an expected error covariance measure. Our results show that for the class of problems studied threshold policies in the error covariance and battery level are optimal.

Notes: Section 3.1 is based on [30], which also considers the case of multiple sensors and imperfect feedback acknowledgements. This work has been extended to LQG control in [35]. Section 3.2 is based on [36]. Subsequent work has additionally studied the problem of transmission scheduling and control with an energy harvesting sensor [37].

Appendix

Proof of Lemma 3.1

We will verify the conditions (CAV*1) and (CAV*2) given in Corollary 7.5.10 of [26], which guarantee the existence of solutions to the Bellman equation for average cost problems with countably infinite state space. Condition (CAV*1) says that there exists a standard policy[5] d such that the recurrent class R_d of the Markov chain induced by d is equal to the entire state space S. Condition (CAV*2) says that given $U > 0$, the set $D_U = \{i \in S | c(i, a) \leq U \text{ for some } a\}$ is finite, where $c(i, a)$ is the cost at each stage when in state i and using action a.

To verify (CAV*1), let d be the policy that always transmits, i.e. $v_k = 1, \forall k$. Let state i of the induced Markov chain correspond to the value $f^i(\bar{P}), i = 0, 1, 2, \ldots$. The state diagram of the induced Markov chain is given in Fig. 3.11, with state space $S = \{0, 1, 2, \ldots\}$.

Let $z = 0$. Then the expected first passage time from state i to state $z = 0$ is

$$\tau_{i,z} = \lambda + 2(1-\lambda)\lambda + 3(1-\lambda)^2\lambda + \cdots = \frac{1}{\lambda} < \infty.$$

---

[5]d is a *standard policy* if there exists a state z such that the expected first passage time $\tau_{i,z}$ from i to z satisfies $\tau_{i,z} < \infty, \forall i \in S$, and the expected first passage cost $c_{i,z}$ from i to z satisfies $c_{i,z} < \infty, \forall i \in S$.

Fig. 3.11 Markov chain for
policy of always transmitting

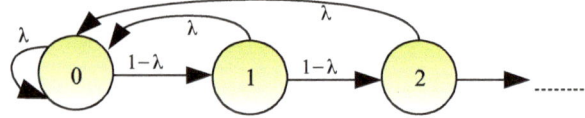

The expected cost of a first passage from state i to state $z = 0$ is

$$
\begin{aligned}
c_{i,z} &= \beta \operatorname{tr} f^i(\bar{P}) + (1-\beta)E + (1-\lambda)c_{(i+1),0} \\
&= \beta \operatorname{tr} f^i(\bar{P}) + (1-\beta)E + (1-\lambda)\left[\beta \operatorname{tr} f^{i+1}(\bar{P}) + (1-\beta)E\right] \\
&\quad + (1-\lambda)^2\left[\beta \operatorname{tr} f^{i+2}(\bar{P}) + (1-\beta)E\right] + \cdots \\
&= \beta \sum_{n=0}^{\infty}(1-\lambda)^n \operatorname{tr} f^{i+n}(\bar{P}) + \frac{(1-\beta)E}{\lambda}.
\end{aligned} \tag{3.33}
$$

For stable A, the infinite series above always converges. To show convergence of the infinite series for unstable A, note that the scenario where the sensor always transmits to the remote estimator, with packet reception probability λ, corresponds to the situation studied in [23, 24]. By computing the stationary probabilities of the Markov chain in Fig. 3.11, we can show that the expected error covariance $\mathbb{E}[P_{k|k}]$ can be written as $\mathbb{E}[P_{k|k}] = \sum_{n=0}^{\infty}(1-\lambda)^n \lambda f^n(\bar{P})$. From the stability results of [23, 24], we know that $\mathbb{E}[P_{k|k}]$ is bounded if and only if $\lambda > 1 - 1/\max_i |\sigma_i(A)|^2$. Thus

$$
\beta \sum_{n=0}^{\infty}(1-\lambda)^n \operatorname{tr} f^{i+n}(\bar{P}) = \frac{\beta}{(1-\lambda)^i \lambda} \sum_{n=0}^{\infty}(1-\lambda)^{i+n} \lambda \operatorname{tr} f^{i+n}(\bar{P}) < \infty
$$

when $\lambda > 1 - 1/\max_i |\sigma_i(A)|^2$.

Hence d is a standard policy. Furthermore, one can see from Fig. 3.11 that the positive recurrent class R_d of the induced Markov chain is equal to S, which verifies (CAV*1).

Since the cost per stage $c(i, a)$ corresponds to $\beta \operatorname{tr} P_{k|k} + (1-\beta)\nu_k E$, condition (CAV*2) can also be easily verified. This thus proves the existence of solutions to the infinite horizon problem (3.9).

Proof of Theorem 3.3

(i) For $B \geq E$, $J_k(P, H, B)$ in (3.31) can be expressed as

$$J_k(P, H, B) = \min\Big\{\operatorname{tr} f(P) + \mathbb{E}[J_{k+1}(f(P), \tilde{H}, \mathfrak{g}(B + \tilde{H}))|H],$$

$$\lambda \operatorname{tr}\bar{P} + (1-\lambda)\operatorname{tr} f(P) + \lambda \mathbb{E}[J_{k+1}(\bar{P}, \tilde{H}, \mathfrak{g}(B - E + \tilde{H}))|H]$$

$$+ (1-\lambda)\mathbb{E}[J_{k+1}(f(P), \tilde{H}, \mathfrak{g}(B - E + \tilde{H}))|H]\Big\},$$

where the two terms in the minimization correspond to the values $v_k = 0$ or $v_k = 1$. Since v_k only takes on the two values 0 and 1, Theorem 3.3 will be proved if we can show that for fixed $B \geq E$ and H, the functions

$$\phi_k(P) \triangleq \operatorname{tr} f(P) + \mathbb{E}[J_{k+1}(f(P), \tilde{H}, \mathfrak{g}(B + \tilde{H}))|H] - \lambda \operatorname{tr}\bar{P} - (1-\lambda)\operatorname{tr} f(P)$$

$$- \lambda \mathbb{E}[J_{k+1}(\bar{P}, \tilde{H}, \mathfrak{g}(B - E + \tilde{H}))|H] - (1-\lambda)\mathbb{E}[J_{k+1}(f(P), \tilde{H}, \mathfrak{g}(B - E + \tilde{H}))|H]$$

$$= \lambda\Big(\operatorname{tr} f(P) - \operatorname{tr}\bar{P} - \mathbb{E}[J_{k+1}(\bar{P}, \tilde{H}, \mathfrak{g}(B - E + \tilde{H}))|H]\Big) + \mathbb{E}[J_{k+1}(f(P), \tilde{H}, \mathfrak{g}(B + \tilde{H}))|H]$$

$$- (1-\lambda)\mathbb{E}[J_{k+1}(f(P), \tilde{H}, \mathfrak{g}(B - E + \tilde{H}))|H]$$

for $k = 1, \ldots, K$, are increasing functions of P. As $\operatorname{tr} f(P)$ is increasing with P by Lemma 3.5, this will be the case if we can show that $\mathbb{E}[J_k(f(P), \tilde{H}, \mathfrak{g}(B + \tilde{H}))|H] - (1-\lambda)\mathbb{E}[J_k(f(P), \tilde{H}, \mathfrak{g}(B - E + \tilde{H}))|H]$ is an increasing function of P for all k.

To prove this using an induction argument, we will in fact prove a slightly stronger statement, namely that

$$J_k(f^n(P), H, B) - (1-\lambda)J_k(f^n(P), H, B') \tag{3.34}$$

is an increasing function of P for all $k \in \{1, \ldots, K + 1\}, n \in \mathbb{N}, H \geq 0, B \geq 0, B' \geq 0$ with $0 \leq B - B' \leq E$, noting that $0 \leq \mathfrak{g}(B + \tilde{H}) - \mathfrak{g}(B - E + \tilde{H}) \leq E$ by Lemma 3.6. In order to show that (3.34) is an increasing function of P, it turns out that we also need to show that

$$J_k(f^n(P), H, B') - J_k(f^n(P), H, B) \tag{3.35}$$

is an increasing function of P for all $k \in \{1, \ldots, K + 1\}, n \in \mathbb{N}, H \geq 0, B \geq 0, B' \geq 0$ with $0 \leq B - B' \leq E$.

As stated before, the proof is by induction. It is clear that (3.34) and (3.35) are increasing functions of P in the case of $k = K + 1$. For $P \geq P'$ and $0 \leq B - B' \leq E$, assume that

$$J_l(f^n(P), H, B) - (1-\lambda)J_l(f^n(P), H, B')$$
$$- J_l(f^n(P'), H, B) + (1-\lambda)J_l(f^n(P'), H, B') \geq 0 \tag{3.36}$$

and

$$J_l(f^n(P), H, B') - J_l(f^n(P), H, B) - J_l(f^n(P'), H, B') + J_l(f^n(P'), H, B) \geq 0$$
$$\tag{3.37}$$

holds for $l = K + 1, K, \ldots, k + 1$.

Let us first show that (3.36) holds for $l = k$. We have

$$
\begin{aligned}
&J_k(f^n(P), H, B) - (1 - \lambda)J_k(f^n(P), H, B') - J_k(f^n(P'), H, B) + (1 - \lambda)J_k(f^n(P'), H, B') \\
&= \min_{v, vE \leq B} \left\{ v\lambda\mathrm{tr}(\bar{P} + (1 - v\lambda)\mathrm{tr}f^{n+1}(P) + v\lambda\mathbb{E}\left[J_{k+1}(\bar{P}, \tilde{H}, \mathfrak{g}(B - vE + \tilde{H}))|H \right] \right. \\
&\quad \left. + (1 - v\lambda)\mathbb{E}\left[J_{k+1}(f^{n+1}(P), \tilde{H}, \mathfrak{g}(B - vE + \tilde{H}))|H \right] \right\} \\
&\quad - (1 - \lambda) \min_{v, vE \leq B'} \left\{ v\lambda\mathrm{tr}\bar{P} + (1 - v\lambda)\mathrm{tr}f^{n+1}(P) \right. \\
&\quad\quad + v\lambda\mathbb{E}\left[J_{k+1}(\bar{P}, \tilde{H}, \mathfrak{g}(B' - vE + \tilde{H}))|H \right] \\
&\quad\quad \left. + (1 - v\lambda)\mathbb{E}\left[J_{k+1}(f^{n+1}(P), \tilde{H}, \mathfrak{g}(B' - vE + \tilde{H}))|H \right] \right\} \\
&\quad - \min_{v, vE \leq B} \left\{ v\lambda\mathrm{tr}\bar{P} + (1 - v\lambda)\mathrm{tr}f^{n+1}(P') + v\lambda\mathbb{E}\left[J_{k+1}(\bar{P}, \tilde{H}, \mathfrak{g}(B - vE + \tilde{H}))|H \right] \right. \\
&\quad\quad \left. + (1 - v\lambda)\mathbb{E}\left[J_{k+1}(f^{n+1}(P'), \tilde{H}, \mathfrak{g}(B - vE + \tilde{H}))|H \right] \right\} \\
&\quad + (1 - \lambda) \min_{v, vE \leq B'} \left\{ v\lambda\mathrm{tr}\bar{P} + (1 - v\lambda)\mathrm{tr}f^{n+1}(P') \right. \\
&\quad\quad + v\lambda\mathbb{E}\left[J_{k+1}(\bar{P}, \tilde{H}, \mathfrak{g}(B' - vE + \tilde{H}))|H \right] \\
&\quad\quad \left. + (1 - v\lambda)\mathbb{E}\left[J_{k+1}(f^{n+1}(P'), \tilde{H}, \mathfrak{g}(B' - vE + \tilde{H}))|H \right] \right\}.
\end{aligned}
$$

If $B \geq E$ and $B' \geq E$, then

$$
\begin{aligned}
&J_k(f^n(P), H, B) - (1 - \lambda)J_k(f^n(P), H, B') \\
&\quad - J_k(f^n(P'), H, B) + (1 - \lambda)J_k(f^n(P'), H, B') \\
&\geq \min_v(1 - v\lambda)\left\{ \lambda\left[\mathrm{tr}f^{n+1}(P) - \mathrm{tr}f^{n+1}(P') \right] \right. \\
&\quad + \mathbb{E}\left[J_{k+1}(f^{n+1}(P), \tilde{H}, \mathfrak{g}(B - vE + \tilde{H}))|H \right] \\
&\quad - (1 - \lambda)\mathbb{E}\left[J_{k+1}(f^{n+1}(P), \tilde{H}, \mathfrak{g}(B' - vE + \tilde{H}))|H \right] \\
&\quad - \mathbb{E}\left[J_{k+1}(f^{n+1}(P'), \tilde{H}, \mathfrak{g}(B - vE + \tilde{H}))|H \right] \\
&\quad \left. + (1 - \lambda)\mathbb{E}\left[J_{k+1}(f^{n+1}(P'), \tilde{H}, \mathfrak{g}(B' - vE + \tilde{H}))|H \right] \right\} \geq 0,
\end{aligned}
$$

where the last inequality holds (for both cases $v^* = 0$ and $v^* = 1$) by Lemma 3.5 and the induction hypothesis (3.36), since $0 \leq \mathfrak{g}(B - vE + \tilde{H}) - \mathfrak{g}(B' - vE + \tilde{H}) \leq E$ when $0 \leq B - B' \leq E$.

If $B < E$ and $B' < E$, then

$$
\begin{aligned}
&J_k(f^n(P), H, B) - (1 - \lambda)J_k(f^n(P), H, B') \\
&\quad - J_k(f^n(P'), H, B) + (1 - \lambda)J_k(f^n(P'), H, B') \\
&= \left\{ \lambda\left[\mathrm{tr}f^{n+1}(P) - \mathrm{tr}f^{n+1}(P') \right] + \mathbb{E}\left[J_{k+1}(f^{n+1}(P), \tilde{H}, \mathfrak{g}(B + \tilde{H}))|H \right] \right.
\end{aligned}
$$

$$- (1 - \lambda)\mathbb{E}\left[J_{k+1}(f^{n+1}(P), \tilde{H}, \mathfrak{g}(B' + \tilde{H}))|H\right]$$

$$- \mathbb{E}\left[J_{k+1}(f^{n+1}(P'), \tilde{H}, \mathfrak{g}(B + \tilde{H}))|H\right]$$

$$+ (1 - \lambda)\mathbb{E}\left[J_{k+1}(f^{n+1}(P'), \tilde{H}, \mathfrak{g}(B' + \tilde{H}))|H\right]\Bigg\} \geq 0,$$

by Lemma 3.5 and the induction hypothesis (3.36).

If $B \geq E$ and $B' < E$, then

$$J_k(f^n(P), H, B) - (1 - \lambda)J_k(f^n(P), H, B')$$
$$- J_k(f^n(P'), H, B) + (1 - \lambda)J_k(f^n(P'), H, B')$$
$$\geq \min_{\nu} \Big\{ \lambda(1 - \nu)\left[\operatorname{tr}f^{n+1}(P) - \operatorname{tr}f^{n+1}(P')\right]$$
$$+ (1 - \nu\lambda)\mathbb{E}\left[J_{k+1}(f^{n+1}(P), \tilde{H}, \mathfrak{g}(B - \nu E + \tilde{H}))|H\right]$$
$$- (1 - \lambda)\mathbb{E}\left[J_{k+1}(f^{n+1}(P), \tilde{H}, \mathfrak{g}(B' + \tilde{H}))|H\right]$$
$$- (1 - \nu\lambda)\mathbb{E}\left[J_{k+1}(f^{n+1}(P'), \tilde{H}, \mathfrak{g}(B - \nu E + \tilde{H}))|H\right]$$
$$+ (1 - \lambda)\mathbb{E}\left[J_{k+1}(f^{n+1}(P'), \tilde{H}, \mathfrak{g}(B' + \tilde{H}))|H\right]\Big\}.$$

In the minimization above, if the optimal $\nu^* = 0$, then $J_k(f^n(P), H, B) - (1 - \lambda)J_k(f^n(P), H, B') - J_k(f^n(P'), H, B) + (1 - \lambda)J_k(f^n(P'), H, B') \geq 0$ by a similar argument as before. If instead $\nu^* = 1$, then we have

$$J_k(f^n(P), H, B) - (1 - \lambda)J_k(f^n(P), H, B')$$
$$- J_k(f^n(P'), H, B) + (1 - \lambda)J_k(f^n(P'), H, B')$$
$$\geq (1 - \lambda)\mathbb{E}\left[J_{k+1}(f^{n+1}(P), \tilde{H}, \mathfrak{g}(B - E + \tilde{H}))|H\right]$$
$$- (1 - \lambda)\mathbb{E}\left[J_{k+1}(f^{n+1}(P), \tilde{H}, \mathfrak{g}(B' + \tilde{H}))|H\right]$$
$$- (1 - \lambda)\mathbb{E}\left[J_{k+1}(f^{n+1}(P'), \tilde{H}, \mathfrak{g}(B - E + \tilde{H}))|H\right]$$
$$+ (1 - \lambda)\mathbb{E}\left[J_{k+1}(f^{n+1}(P'), \tilde{H}, \mathfrak{g}(B' + \tilde{H}))|H\right] \geq 0,$$

where the last inequality now holds by induction hypothesis (3.37), since $0 \leq \mathfrak{g}(B' + \tilde{H}) - \mathfrak{g}(B - E + \tilde{H}) \leq E$ when $0 \leq B - B' \leq E$. This proves that (3.36) holds for $l = k$.

It remains to show that (3.37) holds for $l = k$, i.e. that

$$J_k(f^n(P), H, B') - J_k(f^n(P), H, B) - J_k(f^n(P'), H, B') + J_k(f^n(P'), H, B) \geq 0. \tag{3.38}$$

This can be done using similar arguments as showing that (3.36) holds for $l = k$. If $B \geq E$ and $B' \geq E$, then (3.38) can be shown by making use of the induction hypothesis (3.37). Similarly, (3.38) holds if $B < E$ and $B' < E$. If $B \geq E$ and $B' < E$, then

$$J_k(f^n(P), H, B') - J_k(f^n(P), H, B) - J_k(f^n(P'), H, B') + J_k(f^n(P'), H, B)$$

$$\geq \min_\nu \left\{ \operatorname{tr} f^{n+1}(P) + \mathbb{E}\left[J_{k+1}(f^{n+1}(P), \tilde{H}, \mathfrak{g}(B' + \tilde{H})) | H \right] - (1 - \nu\lambda)\operatorname{tr} f^{n+1}(P) \right.$$

$$- (1 - \nu\lambda)\mathbb{E}\left[J_{k+1}(f^{n+1}(P), \tilde{H}, \mathfrak{g}(B - \nu E + \tilde{H})) | H \right]$$

$$- \operatorname{tr} f^{n+1}(P') - \mathbb{E}\left[J_{k+1}(f^{n+1}(P'), \tilde{H}, \mathfrak{g}(B' + \tilde{H})) | H \right] + (1 - \nu\lambda)\operatorname{tr} f^{n+1}(P')$$

$$\left. + (1 - \nu\lambda)\mathbb{E}\left[J_{k+1}(f^{n+1}(P'), \tilde{H}, \mathfrak{g}(B - \nu E + \tilde{H})) | H \right] \right\}.$$

In the minimization above, if the optimal $\nu^* = 0$, then (3.38) holds by a similar argument as before. If instead $\nu^* = 1$, then (3.38) now holds by Lemma 3.5 and the induction hypothesis (3.36), since $0 \leq \mathfrak{g}(B' + \tilde{H}) - \mathfrak{g}(B - E + \tilde{H}) \leq E$ for $0 \leq B - B' \leq E$. This proves that (3.37) holds for $l = k$.

(ii) This can be proved using similar techniques as in the proof of Theorem 2.4. The details are omitted.

References

1. Y. Xu, J.P. Hespanha, Optimal communication logic in networked control systems, in *Proceedings of the IEEE Conference on Decision and Control*, Paradise Islands, Bahamas (2004), pp. 842–847
2. O.C. Imer, T. Başar, Optimal estimation with limited measurements, in *Proceedings of the IEEE Conference Decision and Control*, Seville, Spain (2005), pp. 1029–1034
3. R. Cogill, S. Lall, J.P. Hespanha, A constant factor approximation algorithm for event-based sampling, in *Proceedings of the American Control Conference*, New York City (2007), pp. 305–311
4. L. Li, M. Lemmon, X. Wang, Event-triggered state estimation in vector linear processes, in *Proceedings of the American Control Conference*, Baltimore, MD (2010), pp. 2138–2143
5. J. Weimer, J. Araújo, K.H. Johansson, Distributed event-triggered estimation in networked systems, in *Proceedings of the IFAC Conference on Analysis and Design of Hybrid Systems*, Eindhoven, Netherlands (2012), pp. 178–185
6. J. Sijs, M. Lazar, Event based state estimation with time synchronous updates. IEEE Trans. Autom. Control **57**(10), 2650–2655 (2012)
7. S. Trimpe, R. D'Andrea, Event-based state estimation with variance-based triggering. IEEE Trans. Autom. Control **59**(12), 3266–3281 (2014)
8. M. Xia, V. Gupta, P.J. Antsaklis, Networked state estimation over a shared communication medium, in *Proceedings of the American Control Conference*, Washington, DC (2013), pp. 4134–4319
9. J. Wu, Q.-S. Jia, K.H. Johansson, L. Shi, Event-based sensor data scheduling: trade-off between communication rate and estimation quality. IEEE Trans. Autom. Control **58**(4), 1041–1046 (2013)

10. D. Han, Y. Mo, J. Wu, B. Sinopoli, L. Shi, Stochastic event-triggered sensor scheduling for remote state estimation, in *Proceedings of the IEEE Conference on Decision and Control*, Florence, Italy (2013), pp. 6079–6084

11. S. Trimpe, Stability analysis of distributed event-based state estimation, in *Proceedings of the IEEE Conference on Decision and Control*, Los Angeles, CA (2014), pp. 2013–2019

12. K.J. Åström, B.M. Bernhardsson, Comparison of Riemann and Lebesgue sampling for first order stochastic systems, in *Proceedings of the IEEE Conference on Decision and Control*, Las Vegas, NV (2002), pp. 2011–2016

13. P. Tabuada, Event-triggered real-time scheduling of stabilizing control tasks. IEEE Trans. Autom. Control **52**(9), 1680–1685 (2007)

14. M. Rabi, K.H. Johansson, Scheduling packets for event-triggered control, in *Proceedings of the European Control Conference*, Budapest, Hungary (2009), pp. 3779–3784

15. C. Ramesh, H. Sandberg, K.H. Johansson, Design of state-based schedulers for a network of control loops. IEEE Trans. Autom. Control **58**(8), 1962–1975 (2012)

16. D.E. Quevedo, V. Gupta, W.-J. Ma, S. Yüksel, Stochastic stability of event-triggered anytime control. IEEE Trans. Autom. Control **59**(12), 3373–3379 (2014)

17. B. Sinopoli, L. Schenato, M. Franceschetti, K. Poolla, M.I. Jordan, S.S. Sastry, Kalman filtering with intermittent observations. IEEE Trans. Autom. Control **49**(9), 1453–1464 (2004)

18. G.M. Lipsa, N.C. Martins, Remote state estimation with communication costs for first-order LTI systems. IEEE Trans. Autom. Control **56**(9), 2013–2025 (2011)

19. A. Nayyar, T. Başar, D. Teneketzis, V.V. Veeravalli, Optimal strategies for communication and remote estimation with an energy harvesting sensor. IEEE Trans. Autom. Control **58**(9), 2246–2260 (2013)

20. M.M. Tentzeris, A. Georgiadis, L. Roselli (eds.), Special issue on energy harvesting and scavenging. Proc. IEEE **102**(11) (2014)

21. S. Ulukus, E. Erkip, P. Grover, K. Huang, O. Simeone, A. Yener, M. Zorzi (eds.), Special issue on wireless communications powered by energy harvesting and wireless energy transfer. IEEE J. Sel. Areas Commun. **33**(3) (2015)

22. B.D.O. Anderson, J.B. Moore, *Optimal Filtering* (Prentice Hall, New Jersey, 1979)

23. Y. Xu, J.P. Hespanha, Estimation under uncontrolled and controlled communications in networked control systems, in *Proceedings of the IEEE Conference on Decision and Control*, Seville, Spain (2005), pp. 842–847

24. L. Schenato, Optimal estimation in networked control systems subject to random delay and packet drop. IEEE Trans. Autom. Control **53**(5), 1311–1317 (2008)

25. D.P. Bertsekas, *Dynamic Programming and Optimal Control*, vol. I, 2nd edn. (Athena Scientific, Belmont, 2000)

26. L.I. Sennott, *Stochastic Dynamic Programming and the Control of Queueing Systems* (Wiley-Interscience, New York, 1999)

27. V. Krishnamurthy, *Partially Observed Markov Decision Processes: From Filtering to Controlled Sensing* (Cambridge University Press, Cambridge, 2016)

28. M.L. Puterman, *Markov Decision Processes: Discrete Stochastic Dynamic Programming* (Wiley-Interscience, New York, 1994)

29. L. Shi, H. Zhang, Scheduling two Gauss-Markov systems: an optimal solution for remote state estimation under bandwidth constraint. IEEE Trans. Signal Process. **60**(4), 2038–2042 (2012)

30. A.S. Leong, S. Dey, D.E. Quevedo, Sensor scheduling in variance based event triggered estimation with packet drops. IEEE Trans. Autom. Control **62**(4), 1880–1895 (2017)

31. A.S. Leong, S. Dey, D.E. Quevedo, On the optimality of threshold policies in event triggered estimation with packet drops, in *Proceedings of the European Control Conference*, Linz, Austria (2015), pp. 921–927

32. Y. Mo, B. Sinopoli, L. Shi, E. Garone, Infinite-horizon sensor scheduling for estimation over lossy networks, in *Proceedings of the IEEE Conference on Decision and Control*, Maui, HI (2012), pp. 3317–3322

33. C.K. Ho, R. Zhang, Optimal energy allocation for wireless communications with energy harvesting constraints. IEEE Trans. Signal Process. **60**(9), 4808–4818 (2012)

34. M. Nourian, A.S. Leong, S. Dey, Optimal energy allocation for Kalman filtering over packet dropping links with imperfect acknowledgments and energy harvesting constraints. IEEE Trans. Autom. Control **59**(8), 2128–2143 (2014)
35. A.S. Leong, D.E. Quevedo, T. Tanaka, S. Dey, A. Ahlén, Event-based transmission scheduling and LQG control over a packet dropping link, in *Proceedings of the IFAC World Congress*, Toulouse, France (2017), pp. 9275–9280
36. A.S. Leong, S. Dey, D.E. Quevedo, Optimal transmission policies for variance based event triggered estimation with an energy harvesting sensor, in *Proceedings of the EUSIPCO*, Budapest, Hungary (2016), pp. 225–229
37. A.S. Leong, S. Dey, D.E. Quevedo, Transmission scheduling for remote state estimation and control with an energy harvesting sensor (2017), submitted for publication

Chapter 4
Optimal Transmission Strategies for Remote State Estimation

Improving system performance and reliability under resource (e.g. energy/power, computation and communication) constraints is one of the important challenges in wireless-based networks. This concern is particularly crucial in industrial applications such as remote sensing and real-time control where a high level of reliability is usually required. As a consequence, it becomes of significant importance to investigate the impact of realistic wireless communication channel models in the area of state estimation and control of networked systems [1]. Two important limitations of wireless communication channels in these problem formulations include (i) limited bandwidth and (ii) information loss or packet loss.

Packet loss has been reviewed and studied in previous chapters. Among the many papers in the area of networked state estimation and control over bandwidth limited channels, we first mention [2], which addresses the minimum data rate required for stability of a linear stochastic system with quantized measurements received through a finite rate channel. This work is extended to the general case of time-varying Markov digital communication channels in [3]. The reader is also referred to the survey [4] and the references therein.

Even though most of the works available in the literature focus on merely one of the two mentioned communication limitations (limited bandwidth or packet loss), some works attempt to address both issues. In particular, the problem of minimum data rates for achieving bounded average state estimation error in linear systems over lossy channels is studied in [5], while the problem of control around a target state trajectory in the case of both signal quantization and packet drops is investigated in [6, 7]. The work in [8] concentrates on designing coding and decoding schemes to remotely estimate the state of a scalar stable stochastic linear system over a communication channel subject to both quantization noise and packet loss.

Similar to [8], the current chapter is concerned with remote state estimation subject to both quantization noise and packet drops. However, rather than considering fixed

© The Author(s) 2018
A.S. Leong et al., *Optimal Control of Energy Resources for State Estimation Over Wireless Channels*, SpringerBriefs in Control, Automation and Robotics, DOI 10.1007/978-3-319-65614-4_4

coding and decoding schemes (as in [8]), we are here interested in choosing optimal transmission policies at the smart sensor that decide between sending the sensor's local state estimates or its local innovations. We present a novel design methodology for optimal transmission policies at a smart sensor to remotely estimate the state of a stable[1] linear stochastic dynamical system. The sensor makes measurements of the process and forms estimates of the state using a local Kalman filter, see Fig. 4.1. The sensor then transmits quantized (using a high-resolution quantizer) information over a packet dropping link to the remote receiver. The sensor decides, at each time instant, whether to transmit a high-rate quantized version of either its local state estimate or its local innovation.

The packet reception probability is generally a function of the length of the packet, such that shorter packets (and hence lower required data rates) may result in higher packet reception probabilities. Since the local innovations process has a smaller covariance, for a fixed packet reception probability, the quantized innovations require less energy to transmit than the quantized state estimates. However, due to the packet dropping link between the sensor and the remote estimator, if there has been a number of successive packet losses, then receiving a quantized state estimate might be more beneficial in reducing the estimation error covariance at the remote estimator than receiving the innovations. Thus, there is a trade-off between whether the sensor should transmit its local state estimates or its local innovations. The objective is to design optimal transmission policies in order to minimize a long-term average (infinite-time horizon) cost function as a linear combination of the receiver's expected estimation error covariance and the energy needed to transmit the packets.

The organization of the chapter is as follows. The system model is given in Sect. 4.1. The augmented state space model at the remote receiver is constructed in Sect. 4.2 and the corresponding Kalman filtering equations are given. Section 4.3 presents optimal transmission policy optimization problems, together with their solutions. For scalar systems, Sect. 4.4 proves the optimality of threshold transmission policies. Numerical simulations are given in Sect. 4.5.

Notation: We let $(\Omega, \mathscr{F}, \mathbb{P})$ denote a complete probability space. Throughout the chapter, the subscript or superscript s is used for the sensor's quantities, and the superscript r is used for the remote estimator's quantities. We say that a matrix $X > 0$ if X is positive definite, and $X \geq 0$ if X is positive semi-definite.

4.1 System Model

A diagram of the system architecture is shown in Fig. 4.1. Detailed descriptions of each part of the systems are given below.

---

[1] We consider stable systems as the state of an unstable system becomes unbounded over time which makes it difficult to quantize its state estimates.

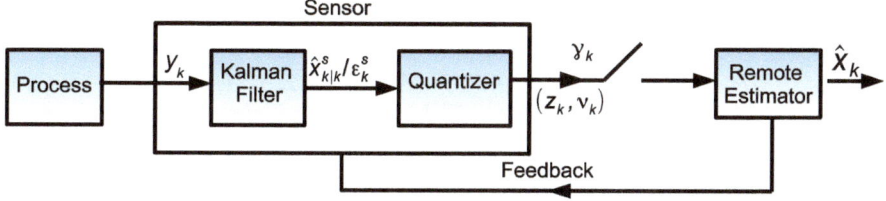

Fig. 4.1 Remote state estimation over a digital packet drop channel

4.1.1 Process Dynamics and Sensor Measurements

We consider a linear dynamical process

$$x_{k+1} = Ax_k + w_k, \tag{4.1}$$

where $x_k \in \mathbb{R}^n$ is the process state at time instant $k \geq 0$, with A being a Schur stable matrix, and $\{w_k : k \geq 0\}$ is a sequence of independent and identically distributed (i.i.d.) Gaussian noises with zero mean and covariance $\Sigma_w \geq 0$.[2] The initial state of the process x_0 is a Gaussian random vector, independent of the process noise sequence $\{w_k : k \geq 0\}$, with mean $\bar{x}_0 := \mathbb{E}[x_0]$ and covariance $P_{x_0} \geq 0$.

The sensor measurements are obtained in the form

$$y_k = Cx_k + v_k, \tag{4.2}$$

where $y_k \in \mathbb{R}^m$ is the vector observation at time instant $k \geq 0$, and $\{v_k\}$ is a sequence of i.i.d. Gaussian noises, independent of both x_0 and $\{w_k\}$, with zero mean and covariance $\Sigma_v > 0$.

4.1.2 Local Kalman Filter at the Smart Sensor

We assume that the sensor has some computational capabilities. In particular, it can run a local Kalman filter to reduce the effects of measurement noise, as in Chap. 3.

Denote the local sensor information at time k by $\mathcal{Y}_k^s := \sigma\{y_t : 0 \leq t \leq k\}$, which is the σ-field generated by the sensor measurements up to time k. We use the convention $\mathcal{Y}_0^s := \{\emptyset, \Omega\}$. The optimal Kalman filtering and prediction estimates of the process state x_k at the sensor are given by $\hat{x}_{k|k}^s = \mathbb{E}[x_k|\mathcal{Y}_k^s]$ and $\hat{x}_{k+1|k}^s = \mathbb{E}[x_{k+1}|\mathcal{Y}_k^s]$, respectively.

We assume that the local Kalman filter at the sensor has reached steady state. The stationary error covariance is defined by $P_s = \lim_{k \to \infty} \mathbb{E}[(x_{k+1} - \hat{x}_{k+1|k}^s)(x_{k+1} -$

---

[2] We use Σ_w and Σ_v rather than Q and R to denote the process and measurement noise covariances, as Q and R will be used for the noise covariances of the augmented state space model in Sect. 4.2.1.

$\hat{x}^s_{k+1|k})^T |\mathscr{Y}^s_k]$, which is the solution of the algebraic Riccati equation (see e.g. [9])

$$P_s = A P_s A^T + \Sigma_w - A P_s C^T (C P_s C^T + \Sigma_v)^{-1} C P_s A^T. \tag{4.3}$$

The Kalman filter equations for $\hat{x}^s_{k|k}$ and $\hat{x}^s_{k+1|k}$ are then given by

$$\hat{x}^s_{k|k} = \hat{x}^s_{k|k-1} + K_f (y_k - C \hat{x}^s_{k|k-1}), \tag{4.4}$$

$$\hat{x}^s_{k+1|k} = A \hat{x}^s_{k|k-1} + K_s (y_k - C \hat{x}^s_{k|k-1}), \tag{4.5}$$

with $\hat{x}^s_{0|-1} := \bar{x}_0$, where $K_f := P_s C^T (C P_s C^T + \Sigma_v)^{-1}$ and $K_s := A K_f$ are the stationary Kalman filtering and prediction gains, respectively. Denote the covariance of the local state estimate by $\Sigma_s := \lim_{k \to \infty} \mathbb{E}[(\hat{x}^s_{k+1|k})(\hat{x}^s_{k+1|k})^T |\mathscr{Y}^s_k]$, which satisfies the stationary Lyapunov equation

$$\Sigma_s = A \Sigma_s A^T + K_s (C P_s C^T + \Sigma_v) K_s^T. \tag{4.6}$$

From (4.4) we can obtain $\lim_{k \to \infty} \mathbb{E}[\hat{x}^s_{k|k}(\hat{x}^s_{k|k})^T] = \Sigma_s + K_f (C P_s C^T + \Sigma_v) K_f^T$.

4.1.3 Coding Alternatives at the Smart Sensor

We define the innovation process at the sensor, $\varepsilon^s_{(\cdot)}$, as

$$\varepsilon^s_k = \hat{x}^s_{k|k} - \hat{x}^s_{k|k-1} = K_f (y_k - C \hat{x}^s_{k|k-1}), \tag{4.7}$$

see (4.4). From (4.7) we can obtain $\lim_{k \to \infty} \mathbb{E}[\varepsilon^s_k (\varepsilon^s_k)^T] = K_f (C P_s C^T + \Sigma_v) K_f^T$. As depicted in Fig. 4.1, the sensor communicates over a digital erasure channel with a remote receiver which utilizes the received data to calculate an estimate of the process state $x_{(\cdot)}$.

This work aims to investigate what data the smart wireless sensor should transmit to the receiver. Motivated by differential Pulse-Code Modulation (PCM) techniques [10, 11], the digital sensor may convey either a vector quantized version of its local estimate, or a vector quantized version of its innovation. Therefore, we may denote the packet sent by the sensor as

$$z_k := \begin{cases} \hat{x}^s_{k|k} + q^x_k, & \text{if } \nu_k = 1 \\ \varepsilon^s_k + q^\varepsilon_k, & \text{if } \nu_k = 0, \end{cases} \tag{4.8}$$

where $\nu_k \in \{0, 1\}$ is a decision variable which is transmitted to the receiver in addition to z_k. The sequence $\{\nu_k\}$ is assumed to be designed at the sensor, though it can also be designed at the remote estimator. In (4.8), $q^x_{(\cdot)}$ and $q^\varepsilon_{(\cdot)}$ are the high-rate quantization noises resulting from encoding $\hat{x}^s_{k|k}$ and ε^s_k, respectively. We note that in this chapter the effects of the quantizer are only modelled via the additional quantization noise term in (4.8). For high-rate quantization, such an approach is quite accurate, since the quantization noises at high rates are approximately uncorrelated with the quantizer inputs [12, 13]. It is also reasonable to assume that the quantization noises, whilst uncorrelated to the inputs, have covariances which are proportional to the input covariances, i.e.

$$
\begin{aligned}
\Sigma_q^x &:= \lim_{k \to \infty} \mathbb{E}[q_k^x (q_k^x)^T] = \alpha_1 \lim_{k \to \infty} \mathbb{E}[\hat{x}_{k|k}^s (\hat{x}_{k|k}^s)^T] \\
\Sigma_q^\varepsilon &:= \lim_{k \to \infty} \mathbb{E}[q_k^\varepsilon (q_k^\varepsilon)^T] = \alpha_0 \lim_{k \to \infty} \mathbb{E}[\varepsilon_k^s (\varepsilon_k^s)^T],
\end{aligned} \tag{4.9}
$$

for given $\alpha_0, \alpha_1 \geq 0$ which depend upon the quantizers and the bit rates used.

To be more specific, consider a vector Gaussian source $\mathfrak{s}$ and a quantizer with $N = 2^n$ quantizer levels, where n is the transmission rate (i.e. the number of bits transmitted per sample). Then the quantization noise covariance of a high-resolution quantizer will be $\Sigma_q \approx \alpha \mathbb{E}[\mathfrak{s}\mathfrak{s}^T]$. For the case of asymptotically optimal lattice vector quantizers with Voronoi cell S_0, we have (see [14]):

$$
\alpha = \frac{M(S_0) V^{2/m} \frac{2}{m} \ln N}{\eta^2} \frac{}{N^{2/m}},
$$

where m represents the dimension of the vector to be quantized, $\eta = \sqrt{1/2}$, $V = \frac{\pi^{m/2}}{\Gamma(m/2+1)}$,

$$
M(S_0) = \frac{\frac{1}{k} \int_{S_0} ||x - y||_2^2 dx}{v(S_0)^{1+2/m}}
$$

is the normalized moment of inertia of S_0, and $v(S_0)$ is the volume of S_0. For $m = 1$, it can be shown that α reduces to $\alpha = \frac{4 \ln N}{3N^2}$. For the case of 'optimal' Lloyd-Max quantizers, we have (see [15])

$$
\alpha \sim \frac{B_m}{N^{2/m}}.
$$

However, the exact values of the constants B_m are not known for dimensions $m \geq 3$. For $m = 1$, we have $\alpha = \frac{\pi \sqrt{3}}{2N^2}$.

In this chapter, we shall focus on a situation wherein the sensor chooses a varying rate of quantization in order to make the traces of the quantization noise covariances Σ_q^x and Σ_q^ε the same. From (4.9), this implies that the data rates n_0 and n_1 for transmitting ε_k^s and $\hat{x}_{k|k}^s$ in the case of the lattice vector quantizer satisfy

$$
\begin{aligned}
\mathrm{Tr}\Sigma_q^x &= \frac{M(S_0) V^{2/m}}{\eta^2} \frac{2n_1 \ln 2/m}{2^{2n_1/m}} \mathrm{Tr}(\Sigma_s + K_f(C P_s C^T + \Sigma_v) K_f^T) \\
&= \frac{M(S_0) V^{2/m}}{\eta^2} \frac{2n_0 \ln 2/m}{2^{2n_0/m}} \mathrm{Tr}(K_f(C P_s C^T + \Sigma_v) K_f^T) = \mathrm{Tr}\Sigma_q^\varepsilon.
\end{aligned}
$$

In the case of the Lloyd-Max quantizer, we have

$$\mathrm{Tr}\Sigma_q^x = \frac{B_m}{2^{2n_1/m}}\mathrm{Tr}(\Sigma_s + K_f(CP_sC^T + \Sigma_v)K_f^T) \tag{4.10}$$

$$= \frac{B_m}{2^{2n_0/m}}\mathrm{Tr}(K_f(CP_sC^T + \Sigma_v)K_f^T) = \mathrm{Tr}\Sigma_q^\varepsilon. \tag{4.11}$$

If the resulting n_0 and n_1 are not integers, their nearest integers will be chosen as the transmission rates. Since the covariance of the local state estimates $\Sigma_s \geq 0$, we have $n_0 \leq n_1$ in the two cases above.

As shown above, the local innovation process has a smaller stationary covariance, and hence a smaller data rate to maintain a given packet reception probability. Therefore, transmitting ε_k^s should require less energy than transmitting $\hat{x}_{k|k}^s$ (see Sect. 4.1.4). However, due to the packet dropping link between the sensor and the remote estimator, if there has been a number of successive packet losses, then receiving $\hat{x}_{k|k}^s$ might be more beneficial in reducing the estimation error covariance at the remote estimator than receiving ε_k^s. Thus, in this model, it is not immediately clear whether the sensor should transmit local estimates $\hat{x}_{k|k}^s$ or innovations ε_k^s. This chapter seeks to elucidate this dilemma in answering how to optimally design the control sequence $\{v_k\}$, using causal information available at the sensor.

4.1.4 Communication Channel

We assume that the forward communication channel between the sensor and the receiver is unreliable, see Fig. 4.1. This channel carries the packets $\{(z_k, v_k) : k \geq 0\}$ and is characterized by the transmission success process $\{\gamma_k : k \geq 0\}$, where $\gamma_k = 1$ refers to successful reception of (z_k, v_k) and $\gamma_k = 0$ quantifies a dropout. Since the decision variable v_k consists of only one bit of information, it can be easily sent along with z_k as a header in the transmitted packet.

In this work we assume that γ_k is a Bernoulli random variable with $\mathbb{P}(\gamma_k = 1) = \lambda = 1 - p$, where $p \in [0, 1]$ is the packet loss probability. The packet loss probability is generally a function of the data rates, such that higher data rates result in higher packet loss probabilities. If p_b is the error probability of sending one bit, then the packet loss probability of sending a packet of n bits will be of the form

$$p = 1 - (1 - p_b)^n, \tag{4.12}$$

where the packet is assumed to be lost if an error occurs in any of its bits (e.g. when there is no channel coding used). We assume that the bit error probability p_b of a wireless communication channel depends on the transmission energy per bit E_b, such that p_b decreases as E_b increases. The bit error probability p_b can be computed for different combinations of channels and digital modulation schemes. For example, in the case of an Additive White Gaussian Noise (AWGN) channel with Binary

Phase-Shift Keying (BPSK) modulation, we have

$$p_b = \mathbf{Q}\left(\sqrt{\frac{2E_b}{N_0}}\right),\tag{4.13}$$

where $N_0/2$ is the noise power spectral density and $\mathbf{Q}(x) := (1/\sqrt{2\pi}) \int_x^\infty e^{-t^2/2} dt = \frac{1}{2}\mathrm{erfc}(\frac{x}{\sqrt{2}})$ is the Q-function [16]. As a consequence of (4.12) and (4.13), to obtain a fixed packet dropout probability, when innovations are sent the transmit energy per bit will be lower than when local estimates are transmitted. In Sect. 4.3 we will further elucidate the situation and allocate power levels accordingly.

The sensor receives an acknowledgment process, such that after the transmission of y_k and before transmitting y_{k+1}, the sensor has access to γ_k.

4.2 Analysis of the System Model

4.2.1 Augmented State Space Model at the Receiver

To analyse the model considered in this chapter, we write the dynamics of the augmented state $\theta_k := [x_k \ \hat{x}^s_{k|k-1}]^T$ which we estimate at the remote receiver as

$$\theta_{k+1} = \mathscr{A}\theta_k + \xi_k,$$

where

$$\mathscr{A} := \begin{bmatrix} A & 0 \\ K_s C & A - K_s C \end{bmatrix} \text{ and } \xi_k := \begin{bmatrix} w_k \\ K_s v_k \end{bmatrix},$$

by (4.1), (4.2) and (4.5). From (4.8), we may write $z_k = v_k(\hat{x}^s_{k|k} + q^x_k) + (1 - v_k)(\varepsilon^s_k + q^\varepsilon_k)$, or

$$z_k = \mathscr{C}(v_k)\theta_k + \zeta_k,$$

where

$$\mathscr{C}(v_k) := [K_f C \ v_k I - K_f C] \text{ and } \zeta_k := K_f v_k + v_k q^x_k + (1 - v_k)q^\varepsilon_k,$$

by (4.2), (4.4) and (4.7) (note that $K_f C$ is a square matrix). We note that $\{\xi_k\}$ and $\{\zeta_k\}$ are zero-mean noise processes. The covariance of the process $\{\xi_k\}$ is

$$Q := \mathbb{E}[\xi_k \xi_k^T] = \begin{bmatrix} \Sigma_w & 0 \\ 0 & K_s \Sigma_v K_s^T \end{bmatrix} \geq 0,$$

while the covariance of the process $\{\zeta_k\}$ is given by

$$R(v_k) := \mathbb{E}[\zeta_k \zeta_k^T] = K_f \Sigma_v K_f^T + v_k^2 \Sigma_q^x + (1 - v_k)^2 \Sigma_q^\varepsilon \geq 0. \qquad (4.14)$$

The matrix S, which models the correlation between the augmented state process noise $\{\xi_k\}$ and the measurement noise $\{\zeta_k\}$, is given by

$$S := \mathbb{E}[\xi_k \zeta_k^T] = \begin{bmatrix} 0 \\ K_s \Sigma_v K_f^T \end{bmatrix}.$$

4.2.2 Kalman Filter at the Receiver

We assume that the receiver knows whether dropouts have occurred or not, and at instances where sensor packets are received, the decision variable v_k is also known. Therefore, the information at the receiver at time k, $\mathscr{Y}_k^r$, is given by the σ-field $\sigma\{\gamma_t, \gamma_t v_t, \gamma_t z_t : 0 \leq t \leq k\}$. We use the convention $\mathscr{Y}_0^r := \{\emptyset, \Omega\}$. At time instant k, the receiver estimates the process state x_k through estimation of the augmented state θ_k based on the information $\mathscr{Y}_{k-1}^r$. We denote the conditional expectation and the associated estimation error covariance of the augmented state as

$$\hat{\theta}_k := \mathbb{E}[\theta_k \mid \mathscr{Y}_{k-1}^r]$$

$$\mathbf{P}_k := \mathbb{E}[(\theta_k - \hat{\theta}_k)(\theta_k - \hat{\theta}_k)^T \mid \mathscr{Y}_{k-1}^r] = \begin{bmatrix} P_k^{1,1} & P_k^{1,2} \\ P_k^{1,2} & P_k^{2,2} \end{bmatrix}. \qquad (4.15)$$

Let $\hat{x}_k^r := \mathbb{E}[x_k \mid \mathscr{Y}_{k-1}^r]$. Then

$$P_k^{1,1} \equiv \mathbb{E}[(x_k - \hat{x}_k^r)(x_k - \hat{x}_k^r)^T \mid \mathscr{Y}_{k-1}^r] \qquad (4.16)$$

is the state estimation error covariance at the receiver at time k. The estimation error covariance $\mathbf{P}_{(\cdot)}$ satisfies the following random Riccati equation of Kalman filtering with correlated process and measurement noises:

$$\mathbf{P}_{k+1} = \mathscr{A}\mathbf{P}_k\mathscr{A}^T + Q - \gamma_k[\mathscr{A}\mathbf{P}_k\mathscr{C}^T(v_k) + S]$$
$$\times [\mathscr{C}(v_k)\mathbf{P}_k\mathscr{C}^T(v_k) + R(v_k)]^{-1}[\mathscr{A}\mathbf{P}_k\mathscr{C}^T(v_k) + S]^T. \qquad (4.17)$$

Note that γ_k appears as a random coefficient in the Riccati equation (4.17). The process $\{v_k\}$ in general is also random, as will become apparent in Sect. 4.3.

Theorem 4.1 *The estimation error covariance $\mathbf{P}_{(\cdot)}$ of the augmented system is of the form*

$$\mathbf{P}_k = \begin{bmatrix} P_k^{1,1} & P_k^{1,1} - P_s \\ P_k^{1,1} - P_s & P_k^{1,1} - P_s \end{bmatrix}, \quad k \geq 0. \tag{4.18}$$

Proof We have by definition $\mathscr{Y}_k^s = \sigma\{y_t : 0 \leq t \leq k\}$ and $\mathscr{Y}_k^r = \sigma\{\gamma_t, \gamma_t v_t, \gamma_t z_t : 0 \leq t \leq k\}$. In addition, let us define the σ-fields: $\mathscr{Y}_k^1 = \sigma\{v_t, y_t : 0 \leq t \leq k\}$, $\mathscr{Y}_k^2 = \sigma\{v_t, z_t : 0 \leq t \leq k\}$, $\mathscr{Y}_k^3 = \sigma\{\gamma_t, v_t, z_t : 0 \leq t \leq k\}$. We have the obvious inclusions $\mathscr{Y}_k^2 \subseteq \mathscr{Y}_k^1$ and $\mathscr{Y}_k^r \subseteq \mathscr{Y}_k^3$. Now $\mathbb{E}[x_k | \mathscr{Y}_{k-1}^s] = \mathbb{E}[x_k | \mathscr{Y}_{k-1}^1]$ since v_{k-1} does not provide any additional information about x_k (because v_k can depend on the error covariance but not the current state). Then we have

$$\mathbb{E}\left[\mathbb{E}[x_k | \mathscr{Y}_{k-1}^s] | \mathscr{Y}_{k-1}^2\right] = \mathbb{E}\left[\mathbb{E}[x_k | \mathscr{Y}_{k-1}^1] | \mathscr{Y}_{k-1}^2\right] = \mathbb{E}[x_k | \mathscr{Y}_{k-1}^2] = \mathbb{E}[x_k | \mathscr{Y}_{k-1}^3],$$

where the second equality is due to the inclusion $\mathscr{Y}_{k-1}^2 \subseteq \mathscr{Y}_{k-1}^1$, and the third equality holds because γ_{k-1} is independent of x_k. Therefore,

$$\mathbb{E}[\hat{x}_k^s | \mathscr{Y}_{k-1}^r] = \mathbb{E}\left[\mathbb{E}[x_k | \mathscr{Y}_{k-1}^s] | \mathscr{Y}_{k-1}^r\right] = \mathbb{E}\left[\mathbb{E}\left[\mathbb{E}[x_k | \mathscr{Y}_{k-1}^s] | \mathscr{Y}_{k-1}^2\right] | \mathscr{Y}_{k-1}^r\right]$$
$$= \mathbb{E}\left[\mathbb{E}[x_k | \mathscr{Y}_{k-1}^3] | \mathscr{Y}_{k-1}^r\right] = \mathbb{E}[x_k | \mathscr{Y}_{k-1}^r] = \hat{x}_k^r,$$

where the second last equality is due to the inclusion $\mathscr{Y}_{k-1}^r \subseteq \mathscr{Y}_{k-1}^3$.

On the other hand,

$$P_k^{2,2} \equiv \mathbb{E}[(\hat{x}_k^s - \mathbb{E}[\hat{x}_k^s | \mathscr{Y}_{k-1}^r])(\hat{x}_k^s - \mathbb{E}[\hat{x}_k^s | \mathscr{Y}_{k-1}^r])^T | \mathscr{Y}_{k-1}^r]$$
$$= \mathbb{E}[(\hat{x}_k^s - \hat{x}_k^r)(\hat{x}_k^s - \hat{x}_k^r)^T | \mathscr{Y}_{k-1}^r]$$
$$= \mathbb{E}\left[\left((x_k - \hat{x}_k^r) - (x_k - \hat{x}_k^s)\right)\left((x_k - \hat{x}_k^r) - (x_k - \hat{x}_k^s)\right)^T | \mathscr{Y}_{k-1}^r\right]$$
$$= P_k^{1,1} + P_s - 2\mathbb{E}[(x_k - \hat{x}_k^r)(x_k - \hat{x}_k^s)^T | \mathscr{Y}_{k-1}^r]. \tag{4.19}$$

We note that $\tilde{x}_k^s := x_k - \hat{x}_k^s$ is orthogonal to $\mathscr{Y}_{k-1}^s$, and hence orthogonal to $\mathscr{Y}_{k-1}^r$. Therefore, $\mathbb{E}[\hat{x}_k^s(\tilde{x}_k^s)^T | \mathscr{Y}_{k-1}^r] = 0$ and $E[\hat{x}_k^r(\tilde{x}_k^s)^T | \mathscr{Y}_{k-1}^r] = 0$, which gives

$$\mathbb{E}[(x_k - \hat{x}_k^r)(x_k - \hat{x}_k^s)^T | \mathscr{Y}_{k-1}^r] = \mathbb{E}[\left((x_k - \hat{x}_k^s) + (\hat{x}_k^s - \hat{x}_k^r)\right)(x_k - \hat{x}_k^s)^T | \mathscr{Y}_{k-1}^r] = P_s.$$

This together with (4.19) implies that $P_k^{2,2} = P_k^{1,1} - P_s$. In a similar way, it can be shown that $P_k^{1,2} = P_k^{1,1} - P_s$. $\quad\square$

Theorem 4.1 is useful in numerical solutions of the stochastic control problems considered in the next section, in that it reduces the dimension (and hence the size) of the state space which needs to be considered.

4.3   Optimal Transmission Policy Problem

Based on the discussion in Sect. 4.1.3, the decision of whether to send the innovation ε_k^s, i.e. set $v_k = 0$, or the state estimate $\hat{x}_{k|k}$, i.e. set $v_k = 1$, will result in bit rates n_0 or n_1, respectively, where $n_0 \leq n_1$. To maintain a fixed packet loss probability p, these bit rates yield different bit error probabilities p_b^0 and p_b^1, where

$$p_b^0 = 1 - (1 - p)^{1/n_0} \geq p_b^1 = 1 - (1 - p)^{1/n_1}$$

by (4.12) and the fact that $n_0 \leq n_1$. The required transmission energies for bit error probabilities p_b^0 and p_b^1 will be denoted by E_b^0 and E_b^1, respectively. Since the transmission energy is a decreasing function of the bit error probability, we have $E_b^0 \leq E_b^1$. For example, in the case of AWGN channel with BPSK modulation, (4.13) implies that

$$E_b^0 = N_0 \left(\text{erfc}^{-1}(2p_b^0) \right)^2 \text{ and } E_b^1 = N_0 \left(\text{erfc}^{-1}(2p_b^1) \right)^2,$$

where $\text{erfc}^{-1}(.)$ is the inverse complementary error function, which is monotonically decreasing.

We define the energy per packet of n bits at time k as

$$J(v_k) = n_{v_k} E_b^{v_k},$$

which depends on the control variable $v_k \in \{0, 1\}$.

We now aim to design optimal transmission policies in order to minimize a linear combination of the trace of the receiver's expected estimation error variance and the amount of energy required at the sensor for sending the packet to the receiver. This optimization problem is formulated as a long-term average (infinite-time horizon) stochastic control problem

$$\min_{\{v_k\}} \limsup_{T \to \infty} \frac{1}{T} \sum_{k=0}^{T-1} \mathbb{E} \left[\beta \text{tr} P_{k+1}^{1,1} + (1-\beta) J(v_k) \big| \{\gamma_l\}_0^{k-1}, \{v_l\}_0^k, P_{x_0} \right], \tag{4.20}$$

where $\beta \in [0, 1]$ is the weight, and $P_{k+1}^{1,1}$ is the submatrix of $\mathbf{P}_{k+1}$ (see (4.15) and (4.16)) obtained from the Riccati equation (4.17). The problem (4.20) may be rewritten as

$$\min_{\{v_k\}} \limsup_{T \to \infty} \frac{1}{T} \sum_{k=0}^{T-1} \mathbb{E} \left[\beta \text{tr} P_{k+1}^{1,1} + (1-\beta) J(v_k) \big| \mathbf{P}_k, v_k \right] \tag{4.21}$$

due to the fact that $\mathbf{P}_k$ is a deterministic function of $\{\gamma_l\}_{l=0}^{k-1}$, $\{v_l\}_{l=0}^{k-1}$, and P_{x_0}. Denote

$$\mathscr{L}(\mathbf{P}, \gamma, \nu) := \mathscr{A}\mathbf{P}\mathscr{A}^T + Q - \gamma[\mathscr{A}\mathbf{P}\mathscr{C}^T(\nu) + S]$$
$$[\mathscr{C}(\nu)\mathbf{P}\mathscr{C}^T(\nu) + R(\nu)]^{-1}[\mathscr{A}\mathbf{P}\mathscr{C}^T(\nu) + S]^T$$
$$\equiv \begin{bmatrix} \mathscr{L}^{1,1}(\mathbf{P}, \gamma, \nu) & \mathscr{L}^{1,1}(\mathbf{P}, \gamma, \nu) - P_s \\ \mathscr{L}^{1,1}(\mathbf{P}, \gamma, \nu) - P_s & \mathscr{L}^{1,1}(\mathbf{P}, \gamma, \nu) - P_s \end{bmatrix} \tag{4.22}$$

as the random Riccati equation operator (which has the form (4.22) by Theorem 4.1), where the matrices $\mathscr{A}$, Q, $\mathscr{C}$, S and R are given in Sect. 4.2.1.

Theorem 4.2 *Independent of the initial estimation error variance P_{x_0}, the value of problem (4.21) is given by ρ, which is the solution of the average cost optimality (Bellman) equation*

$$\rho + V(\mathbf{P}) = \min_{\nu \in \{0,1\}} \left(\mathbb{E}\big[\beta tr\mathscr{L}^{1,1}(\mathbf{P}, \gamma, \nu) + (1-\beta)J(\nu)\big|\mathbf{P}, \nu\big] + \mathbb{E}\big[V(\mathscr{L}(\mathbf{P}, \gamma, \nu))\big|\mathbf{P}, \nu\big]\right),$$
$$\tag{4.23}$$

where V is the relative value function.

Proof The proof uses similar techniques as in the proof of Theorem 2.3.   □

In (4.23), the term $\mathbb{E}\big[\mathscr{L}^{1,1}(\mathbf{P}, \gamma, \nu)\big|\mathbf{P}, \nu\big]$ is the submatrix (similar to (4.15)) of

$$\mathbb{E}\big[\mathscr{L}(\mathbf{P}, \gamma, \nu)\big|\mathbf{P}, \nu\big] = APA^T + Q - (1-p)[APC^T(\nu) + S]$$
$$\times [C(\nu)\mathbf{P}C^T(\nu) + R(\nu)]^{-1}[APC^T(\nu) + S]^T, \tag{4.24}$$

where p is the packet loss probability of the forward erasure communication channel given in Sect. 4.1.4. For a given $\mathbf{P}$, the stationary solution to the stochastic control problem (4.21) is then given by the ν that solves the average cost Bellman equation (4.23), and can be found by the use of the relative value iteration algorithm (see Sect. 4.4).

4.4   Structural Results on Optimal Transmission Policies for Scalar Systems

This section presents structural results on the optimal transmission policies for scalar systems (where we will set $A = a \in \mathbb{R}$, $C = 1$, $\Sigma_w = \sigma_w^2$, $\Sigma_v = \sigma_v^2$, $\Sigma_q = \sigma_q^2$). The idea is to apply the submodularity concept (see [17, 18]) to the Bellman equation (4.23), to show that the optimal policy $\nu^*(\cdot)$ is monotonically increasing with respect to the receiver's state estimation error variance $\mathbf{P}^{1,1}$. This monotonicity then implies a threshold structure since the control space has only the two elements in $\{0, 1\}$.

Definition 4.1 ([17] *after* [18]) A function $F(x, y) : X \times Y \to S$ is submodular in (x, y) if $F(x_1, y_1) + F(x_2, y_2) \le F(x_1, y_2) + F(x_2, y_1)$ for all $x_1, x_2 \in X$ and $y_1, y_2 \in Y$ such that $x_1 \ge x_2$ and $y_1 \ge y_2$.   □

It is important to note that submodularity is a sufficient condition for optimality of monotone increasing policies. Specifically, if $F(x, y)$ defined above is submodular in (x, y), then $y(x) = \arg\min_y F(x, y)$ is non-decreasing in x [18].

Define the class of matrices $\mathbb{S}$ as

$$\mathbb{S} := \left\{ \mathbf{P} = \begin{bmatrix} P & P - P_s \\ P - P_s & P - P_s \end{bmatrix} : P \geq P_s \right\}. \tag{4.25}$$

We define an ordering $\geq$ for matrices in class $\mathbb{S}$ as $\mathbf{P}_1 \geq \mathbf{P}_2$ if $\mathbf{P}_1 - \mathbf{P}_2$ is positive semi-definite. It is evident that for $\mathbf{P}_1, \mathbf{P}_2 \in \mathbb{S}$, we have $\mathbf{P}_1 \geq \mathbf{P}_2$ if and only if $P_1^{1,1} \geq P_2^{1,1}$. We also define the mapping $F : \mathbb{S} \times \{0, 1\} \to \mathbb{S}$ as

$$F(\mathbf{P}, \nu) = \mathscr{A}\mathbf{P}\mathscr{A}^T + Q - (1 - p)[\mathscr{A}\mathbf{P}\mathscr{C}^T(\nu) + S]$$
$$[\mathscr{C}(\nu)\mathbf{P}\mathscr{C}^T(\nu) + R]^{-1}[\mathscr{A}\mathbf{P}\mathscr{C}^T(\nu) + S]^T,$$

in view of the instantaneous cost $\mathbb{E}\big[\mathscr{L}(\mathbf{P}, \gamma, \nu)|\mathbf{P}, \nu\big]$ in (4.24). Note that in the scalar case R can be made independent of ν_k (see (4.14)), since in the scalar case we can always achieve $\Sigma_q^x = \Sigma_q^\varepsilon$.

Lemma 4.1 *For scalar systems, the function $F(\mathbf{P}, \nu)$ is submodular in $(\mathbf{P}, \nu)$, i.e. for $\mathbf{P}_1, \mathbf{P}_2 \in \mathbb{S}$ such that $\mathbf{P}_1 \geq \mathbf{P}_2$, we have*

$$F(\mathbf{P}_1, 1) + F(\mathbf{P}_2, 0) \leq F(\mathbf{P}_1, 0) + F(\mathbf{P}_2, 1). \tag{4.26}$$

Proof We will show that

$$F^{1,1}(\mathbf{P}_1, 1) + F^{1,1}(\mathbf{P}_2, 0) \leq F^{1,1}(\mathbf{P}_1, 0) + F^{1,1}(\mathbf{P}_2, 1), \tag{4.27}$$

where $F^{1,1}(\cdot, \cdot)$ is the $(1, 1)$ entry of $F(\cdot, \cdot)$. This will then imply (4.26). Let

$$\mathbf{P} := \begin{bmatrix} P^{1,1} & P^{1,1} - P_s \\ P^{1,1} - P_s & P^{1,1} - P_s \end{bmatrix}$$

where $P^{1,1} \geq P_s$, which implies that $\mathbf{P} \in \mathbb{S}$. First, note that

$$F^{1,1}(\mathbf{P}, 0) = a^2 P^{1,1} + \sigma_w^2 - (1 - p)\frac{a^2 K_f^2 P_s^2}{K_f^2 P_s + R}$$

and

$$F^{1,1}(\mathbf{P}, 1) = a^2 P^{1,1} + \sigma_w^2 - (1 - p)\frac{a^2\big((P^{1,1} - P_s) + K_f P_s\big)^2}{(P^{1,1} - P_s) + K_f^2 P_s + R}.$$

We denote

$$g(x) := \frac{\big((x - P_s) + K_f P_s\big)^2}{(x - P_s) + K_f^2 P_s + R}, \quad x \geq P_s.$$

Let $\mathbf{P}_1, \mathbf{P}_2 \in \mathbb{S}$ be such that $\mathbf{P}_1 \geq \mathbf{P}_2$. Then the inequality (4.27) is equivalent to

$$a^2(P_1^{1,1} - P_2^{1,1}) - (1 - p)a^2\big(g(P_1^{1,1}) - g(P_2^{1,1})\big) \leq a^2(P_1^{1,1} - P_2^{1,1}). \tag{4.28}$$

On the other hand, the derivative $g'(x)$ satisfies

$$g'(x) = \frac{2\big((x - P_s) + K_f P_s\big)\big((x - P_s) + K_f^2 P_s + R\big)}{\big((x - P_s) + K_f^2 P_s + R\big)^2} - \frac{\big((x - P_s) + K_f P_s\big)^2}{\big((x - P_s) + K_f^2 P_s + R\big)^2}. \tag{4.29}$$

In the case $x = P_s - K_f P_s$ where either $P_s = 0$ or $\sigma_v^2 = 0$, (4.29) yields $g'(x) = 0$. Otherwise, dividing the numerator of the right-hand side of (4.29) by the positive value $(x - P_s) + K_f P_s$, one obtains

$$2\big((x - P_s) + K_f^2 P_s + R\big) - \big((x - P_s) + K_f P_s\big) = x - P_s + 2K_f^2 P_s + 2R - K_f P_s$$
$$= x - P_s + 2K_f^2(P_s + \sigma_v^2) + 2\sigma_q^2 - P_s^2/(P_s + \sigma_v^2)$$

by the fact that $R = K_f^2 \sigma_v^2 + \sigma_q^2$. Since $K_f = P_s/(P_s + \sigma_v^2)$, we have

$$x - P_s + 2K_f^2(P_s + \sigma_v^2) + 2\sigma_q^2 - P_s^2/(P_s + \sigma_v^2) = x - P_s + P_s^2/(P_s + \sigma_v^2) + 2\sigma_q^2 \geq 0$$

Therefore, $g'(x) \geq 0$ for $x \geq P_s$.

Since $g(x)$ is an increasing function of $x \geq P_s$, and $P_1^{1,1} \geq P_2^{1,1} \geq P_s$, the inequality (4.28) is valid. This gives (4.27), and thus

$$F(\mathbf{P}_1, 1) - F(\mathbf{P}_2, 1) \leq F(\mathbf{P}_1, 0) - F(\mathbf{P}_2, 0),$$

based on Theorem 4.1 and the fact that for $\mathbf{P}_1, \mathbf{P}_2 \in \mathbb{S}$, we have $\mathbf{P}_1 \geq \mathbf{P}_2$ if and only if $P_1^{1,1} \geq P_2^{1,1}$. $\square$

We now present the relative value iteration algorithm to solve the Bellman equation (4.23). It is used to derive structural results for the optimal transmission policy. First, we consider the Bellman equation for the finite T-horizon stochastic control problem:

$$V_t(\mathbf{P}) = \min_{v \in \{0,1\}} \Big(\mathbb{E}\big[\beta \mathscr{L}^{1,1}(\mathbf{P}, \gamma, v) + (1 - \beta)J(v)|\mathbf{P}, v\big]$$
$$+ \mathbb{E}\big[V_{t+1}\big(\mathscr{L}(\mathbf{P}, \gamma, v)\big)|\mathbf{P}, v\big]\Big), \quad 0 \leq t \leq T - 1 \tag{4.30}$$

with terminal condition $V_T(\mathbf{P}) = 0$, where T is large. We now define the function

$$H_t(\cdot) := V_t(\cdot) - V_t(\mathbf{P}_f), \quad 0 \le t \le T \tag{4.31}$$

where $\mathbf{P}_f \neq \mathbf{P}_0$ is fixed. We then have the following relative value iteration algorithm recursion:

$$
\begin{aligned}
H_t(\mathbf{P}) = \min_{v \in \{0,1\}} &\left(\mathbb{E}\big[\beta \mathscr{L}^{1,1}(\mathbf{P}, \gamma, v) + (1-\beta)J(v)\big|\mathbf{P}, v\big] \right. \\
&\left. + \mathbb{E}\big[V_{t+1}(\mathscr{L}(\mathbf{P}, \gamma, v))|\mathbf{P}, v\big] \right) \\
- \min_{v \in \{0,1\}} &\left(\mathbb{E}\big[\beta \mathscr{L}^{1,1}(\mathbf{P}, \gamma, v) + (1-\beta)J(v)\big|\mathbf{P} = \mathbf{P}_f, v\big] \right. \\
&\left. + \mathbb{E}\big[V_{t+1}(\mathscr{L}(\mathbf{P}, \gamma, v))|\mathbf{P} = \mathbf{P}_f, v\big] \right)
\end{aligned}
\tag{4.32}
$$

for $0 \le t \le T - 1$. It can be shown that the relative value iteration recursion (4.32) converges to the optimal solution ρ of the infinite-time horizon average cost Bellman equation (4.23) such that $\rho \approx H_0(\mathbf{P}_0)$ (see the discussion on p. 391 of [19]).

Theorem 4.3 *For scalar systems, the optimal transmission policy is a threshold policy with respect to the receiver's state estimation error variance $P^{1,1}$ (and hence in the augmented state estimation error covariance $\mathbf{P}$), i.e.*

$$v^*(\mathbf{P}) = \begin{cases} 0, & \text{if } P_k^{1,1} \le \phi^* \\ 1, & \text{otherwise,} \end{cases} \tag{4.33}$$

where ϕ^* is the threshold.

Proof Based on the relative value iteration (4.32), define

$$
\begin{aligned}
L_t(\mathbf{P}, v) &= \mathbb{E}\big[\beta \mathscr{L}^{1,1}(\mathbf{P}, \gamma, v) + (1-\beta)J(v)\big|\mathbf{P}, v\big] + \mathbb{E}\big[V_{t+1}(\mathscr{L}(\mathbf{P}, \gamma, v))|\mathbf{P}, v\big] \\
&:= L_t^{(1)}(\mathbf{P}, v) + L_t^{(2)}(\mathbf{P}, v)
\end{aligned}
$$

for $0 \le t \le T - 1$. We will next employ the submodularity concept.

Submodularity of $L_t^{(1)}(\mathbf{P}, v)$: Lemma 4.1 implies that $F(\mathbf{P}, v) = \mathbb{E}\big[\mathscr{L}(\mathbf{P}, \gamma, v)\big|\mathbf{P}, v\big]$, and hence $\big(F(P, v)\big)^{1,1} = \mathbb{E}\big[\mathscr{L}^{1,1}(\mathbf{P}, \gamma, v)\big|\mathbf{P}, v\big]$, are submodular in $(\mathbf{P}, v)$. It is evident that $\mathbb{E}\big[J(v)\big|v\big]$ is also submodular in $(\mathbf{P}, v)$, since it is independent of $\mathbf{P}$. Therefore, their linear combination $L_t^{(1)}(\mathbf{P}, v)$ is submodular in $(\mathbf{P}, v)$.

Submodularity of $L_t^{(2)}(\mathbf{P}, v)$: First we note that both $\mathscr{L}(\mathbf{P}, \gamma, 0) = \mathbb{E}\big[\mathscr{L}(\mathbf{P}, \gamma, v)\big|\mathbf{P}, v = 0\big]$ and $\mathscr{L}(\mathbf{P}, \gamma, 1) = \mathbb{E}\big[\mathscr{L}(\mathbf{P}, \gamma, v)\big|\mathbf{P}, v = 1\big]$ given in (4.24) are concave[3] and non-decreasing functions in $\mathbf{P}$ (see Lemmas 1 and 2 in [20]). This implies that

$$
\begin{aligned}
\mathbb{E}\big[\beta \mathscr{L}^{1,1}(\mathbf{P}, \gamma, v) + (1-\beta)J(v)\big|\mathbf{P}, v = 0\big], \\
\mathbb{E}\big[\beta \mathscr{L}^{1,1}(\mathbf{P}, \gamma, v) + (1-\beta)J(v)\big|\mathbf{P}, v = 1\big],
\end{aligned}
$$

---

[3]The proof of concavity is based on the fact that a function $f(x)$ is concave in x if and only if $f(x_0 + th)$ is concave in the scalar t for all x_0 and h.

and therefore

$$\min_{v\in\{0,1\}} \left(\mathbb{E}\left[\beta \mathcal{L}^{1,1}(\mathbf{P}, \gamma, v) + (1-\beta)J(v) \big| \mathbf{P}, v \right] \right),$$

are concave and non-decreasing functions of $\mathbf{P}$ (note that the expectation operator preserves concavity). By induction and the fact that the composition of two non-decreasing concave functions is itself concave and non-decreasing, one can show that the value function $V_t(\mathbf{P})$ in (4.30) is a concave and non-decreasing function of $\mathbf{P}$. But the composition of a non-decreasing concave function $V_t(\cdot)$ with a monotonic submodular function $\mathcal{L}(\cdot, \gamma, v)$ is submodular (see part (c) of Proposition 2.3.5 in [21]). Therefore, $L_t^{(2)}(\mathbf{P}, v) = \mathbb{E}\left[V_{t+1}\left(\mathcal{L}(\mathbf{P}, \gamma, v)\right) | \mathbf{P}, v \right]$ is submodular in $(\mathbf{P}, v)$.

Submodularity of $L_t(\mathbf{P}, v)$: The sum of two submodular functions $L_t(\mathbf{P}, v) = L_t^{(1)}(\mathbf{P}, v) + L_t^{(2)}(\mathbf{P}, v)$ is also submodular.

As a result of submodularity,

$$\arg\min_{v\in\{0,1\}} L_t(\mathbf{P}, v), \qquad 0 \le t \le T-1$$

is non-decreasing in $\mathbf{P}$ (see [18]), and hence non-decreasing in $P^{1,1}$. This monotonicity implies the threshold structure (4.33), since the control space $\{0, 1\}$ has only two elements. $\square$

Similar to the structural results obtained in the preceding chapters, knowing that the optimal policy is a threshold policy simplifies real-time implementation, and additionally allows one to derive specialized algorithms which provide computational savings when computing the optimal threshold ϕ^* numerically, see [22].

4.5   Numerical Studies

We present here numerical results for a scalar model with parameters $a = 0.95$, $\sigma_w^2 = 0.25$, $\sigma_v^2 = 0.01$ and $P_{x_0} = 1$ in (4.1) and (4.2). These values give $P_s = 0.26$, $K_s = 0.91$, $K_f = 0.96$ and $\Sigma_s = 2.30$, see Sect. 4.1.2. We take $\sigma_q^2 = \Sigma_q^x = \Sigma_q^\varepsilon = 0.01$ in (4.9) together with a Lloyd-Max quantizer, which yields $n_0 = 3$ and $n_1 = 5$ by (4.11). In the simulation results, an AWGN channel with BPSK modulation is assumed where $N_0 = 0.01$ in (4.13).

First, let the packet error probability p in (4.12) be equal to 0.2. This gives $p_b^0 = 0.07$ and $p_b^1 = 0.04$, and hence, energy per bit levels of $E_b^0 = 0.21$ and $E_b^1 = 0.29$, see Sect. 4.3.

In Fig. 4.2, we plot the expected estimation error covariance versus the packet error probabilities. We let the transmission policies $\{v_k\}$ be fixed either to zero (sending innovations) or one (sending state estimates). On the other hand, Fig. 4.3 presents the packet transmission energy $J(v)$ (in milliwatt hour (mWh)) defined in Sect. 4.3 versus the packet error probabilities. Figures 4.2 and 4.3 show that transmitting local

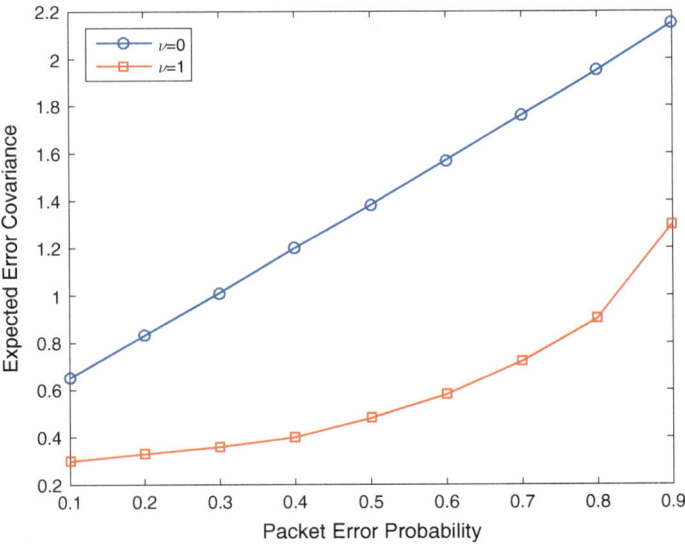

Fig. 4.2 Expected estimation error covariance versus the packet error probabilities for the two cases $\nu = 0$ and $\nu = 1$

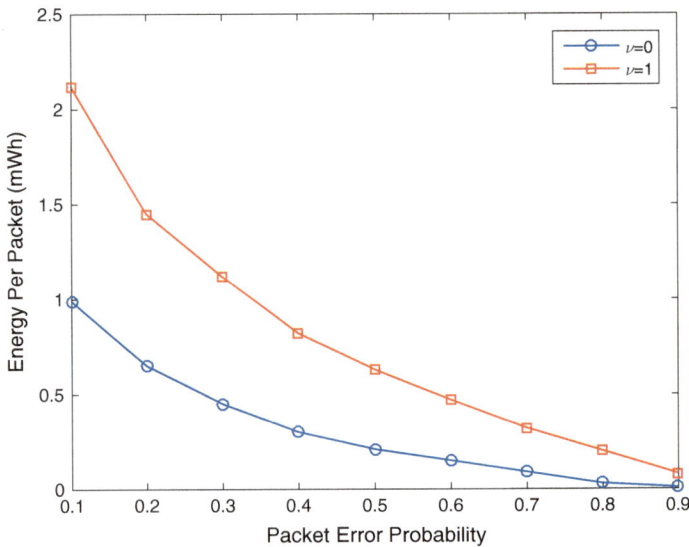

Fig. 4.3 Transmission energy per packet versus the packet error probabilities for the two cases $\nu = 0$ and $\nu = 1$

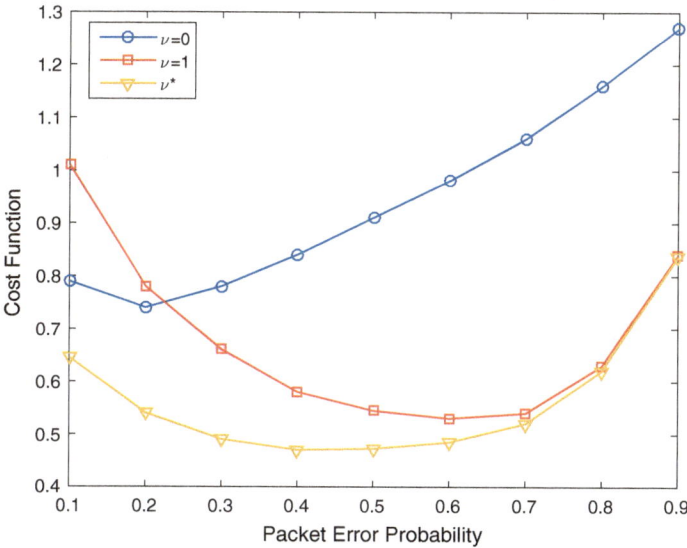

Fig. 4.4 Performance versus packet error probabilities

estimates gives smaller error covariance, but also requires more transmit energy, than transmitting local innovations. This fact motivates the optimization formulation (4.20).

We now set the weight β in problem (4.20) to 0.6. The discretized equation of the relative value iteration algorithm (4.32) is used for the numerical computation of the optimal transmission policy. In solving the Bellman equation (4.23) we use 40 discretization points for the state estimate error variance $P_k^{1,1}$ in the range of [0,1]. In Fig. 4.4 we plot the cost function consisting of the linear combination of the receiver's expected estimation error covariance and the energy needed to transmit the packets, versus the packet loss probability $p \in [0.1, 0.9]$ for the cases of (i) fixed transmission policy $\nu = 0$, (ii) fixed transmission policy $\nu = 1$ and (iii) optimal transmission policy ν^*. We observe that for small packet loss probabilities, sending innovations ($\nu = 0$) is better than sending the state estimates ($\nu = 1$). On the other hand, for large packet loss probabilities sending the state estimates gives better performance than sending the innovations, due to the poor estimation performance when sending innovations when the packet loss probability is high.

4.6 Conclusion

This chapter presents a design methodology for remote state estimation of a stable linear dynamical system, subject to packet dropouts. The key novelty of this formulation is that the smart sensor decides, at each discrete-time instant, whether to transmit either its local state estimate or its local innovation. It is shown how to design optimal transmission policies in order to minimize a long-term average (infinite-time horizon) cost function as a linear combination of the receiver's expected estimation error covariance and the energy needed to transmit the packets. For scalar systems, the optimality of a threshold policy is proved.

Notes: This chapter is based on [22], which also considers the case of imperfect feedback acknowledgements, and a stochastic gradient algorithm for computing the optimal threshold.

References

1. J. Baillieul, P.J. Antsaklis, Control and communication challenges in networked real-time systems. Proc. IEEE **95**(1), 9–28 (2007)
2. G.N. Nair, R.J. Evans, Stabilizability of stochastic linear systems with finite feedback data rates. SIAM J. Control Optim. **43**(2), 413–436 (2004)
3. P. Minero, L. Coviello, M. Franceschetti, Stabilization over Markov feedback channels: the general case. IEEE Trans. Autom. Control **58**(2), 349–362 (2013)
4. G.N. Nair, F. Fagnani, S. Zampieri, R.J. Evans, Feedback control under data rate constraints: an overview. Proc. IEEE **95**(1), 108–137 (2007)
5. K. You, L. Xie, Minimum data rate for mean square stabilization of discrete LTI systems over lossy channels. IEEE Trans. Autom. Control **55**(10), 2373–2378 (2010)
6. M. Trivellato, N. Benvenuto, State control in networked control systems under packet drops and limited transmission bandwidth. IEEE Trans. Commun. **58**(2), 611–622 (2010)
7. D.E. Quevedo, A. Ahlén, D. Nesic, Packetized predictive control of stochastic systems over bit-rate limited channels with packet loss. IEEE Trans. Autom. Control **56**(12), 2854–2868 (2011)
8. S. Dey, A. Chiuso, L. Schenato, Remote estimation with noisy measurements subject to packet loss and quantization noise. IEEE Trans. Control Netw. Syst. **1**(3), 204–217 (2014)
9. B.D.O. Anderson, J.B. Moore, *Optimal Filtering* (Prentice Hall, New Jersey, 1979)
10. D.E. Quevedo, A. Ahlén, J. Østergaard, G.C. Goodwin, Innovations-based state estimation with wireless sensor networks, in *Proceedings of the ECC*, Budapest, Hungary (2009), pp. 4858–4864
11. N.S. Jayant, P. Noll, *Digital Coding of Waveforms* (Prentice Hall, Englewood Cliffs, 1984)
12. D. Marco, D.L. Neuhoff, The validity of the additive noise model for uniform scalar quantizers. IEEE Trans. Inf. Theory **51**(5), 1739–1755 (2005)
13. V.K. Goyal, High-rate transform coding: How high is high, and does it matter? in *Proceedings of the ISIT*, Sorrento, Italy (2000), p. 207
14. P.W. Moo, Asymptotic analysis of lattice-based quantization. Ph.D. dissertation, University of Michigan (1998)
15. A. Gersho, Asymptotically optimal block quantization. IEEE Trans. Inf. Theory **25**(4), 373–380 (1979)
16. J.G. Proakis, *Digital Communications*, 4th edn. (McGraw-Hill, New York, 2001)
17. M.H. Ngo, V. Krishnamurthy, Optimality of threshold policies for transmission scheduling in correlated fading channels. IEEE Trans. Commun. **57**(8), 2474–2483 (2009)
18. D.M. Topkis, *Supermodularity and Complementarity* (Princeton University Press, Princeton, 2001)
19. D.P. Bertsekas, *Dynamic Programming and Optimal Control*, vol. I, 2nd edn. (Athena Scientific, Belmont, 2000)
20. V. Gupta, T.H. Chung, B. Hassibi, R.M. Murray, On a stochastic sensor selection algorithm with applications in sensor scheduling and sensor coverage. Automatica **42**(2), 251–260 (2006)

21. D. Simchi-Levi, X. Chen, J. Bramel, *The Logic of Logistics: Theory, Algorithms, and Applications for Logistics and Supply Chain Management* (Springer, Berlin, 2004)
22. M. Nourian, A.S. Leong, S. Dey, D.E. Quevedo, An optimal transmission strategy for Kalman filtering over packet dropping links with imperfect acknowledgements. IEEE Trans. Control Netw. Syst. **1**(3), 259–271 (2014)

Chapter 5
Remote State Estimation in Multi-hop Networks

This chapter focuses on remote state estimation problems when using multiple sensors and multi-hop networks. In communications, performance benefits can be obtained if one adopts advanced communication techniques such as network coding [1, 2], relays [3] and rerouting [4]. Here, we will show how these concepts can be employed in remote state estimation. We first consider, in Sect. 5.1, a setup where sensors can transmit both directly to the remote estimator or via intermediate relays. We consider different operations that the relay can perform such as forwarding of transmissions or network coding operations, and optimize over the relay operations and transmission powers. Next, we consider in Sect. 5.2 the problem of reconfiguring the topology of (or rerouting) a multi-hop network, in order to respond to time variations in the wireless channel conditions. Optimal and suboptimal methods for reconfiguring the network are presented and their performances compared.

Notation: We define $\mathrm{col}(X_1, \ldots, X_n) \triangleq [\, X_1^T \ldots X_n^T \,]^T$ to be the matrix formed by stacking the matrices $X_1, \ldots, X_n$ on top of each other, and $\mathrm{diag}(X_1, \ldots, X_n)$ to be the block diagonal matrix with $X_1, \ldots, X_n$ being the diagonal blocks. We say that a matrix $X > 0$ if X is positive definite, and $X \geq 0$ if X is positive semi-definite.

5.1 Kalman Filtering over Fading Channels with Relays

5.1.1 Background

In digital communications, channel coding is often used to improve the quality of transmissions over unreliable channels. The concept of network coding [1, 2], where in a network with many nodes, throughput can be increased by allowing intermediate nodes to perform simple operations (such as linear transformations [1]) on its received

© The Author(s) 2018   85
A.S. Leong et al., *Optimal Control of Energy Resources for State Estimation Over Wireless Channels*, SpringerBriefs in Control, Automation and Robotics, DOI 10.1007/978-3-319-65614-4_5

information, has attracted much attention in recent years. Kalman filtering with power control and coding was considered in [5, 6]. The work [6] included a study of network coding, where one could choose to utilize a relay to perform a network coding operation and the energy trade-offs involved. The use of relays in combating the effects of fading and increasing channel capacity has been extensively studied in wireless communications, see e.g. [3, 7]. Indeed, cooperative communications via the use of relays have been identified as one of the key enabling technologies for fifth-generation (5G) mobile networks [8]. The use of a relay in control has been studied in [9], which showed that for the case of a single sensor the stability region for stabilizing an unstable LTI plant can be enlarged in some situations, and also in applications towards control of unmanned aerial vehicles [10].

In this section, we will study remote estimation using relays and investigate what information the individual relays should send to the gateway/fusion centre. In a related setup considered in [6], the relay could only perform network coding that linearly combines two of the sensor transmissions using an XOR operation [1]. Here, we allow for the possibility of the relay combining multiple sensor transmissions using XOR operations [11], as well as the possibility of the relay forwarding the sensors' transmissions, which has the potential to give better performance.

5.1.2 System Model

The process is a discrete-time linear system

$$x_{k+1} = Ax_k + w_k, \tag{5.1}$$

where $x_k \in \mathbb{R}^n$ and $\{w_k\}$ is i.i.d. Gaussian with zero mean and covariance matrix $Q > 0$. The process is observed by M sensors with measurements

$$y_{i,k} = C_i x_k + v_{i,k}, \quad i = 1, \ldots, M, \tag{5.2}$$

where $y_{i,k} \in \mathbb{R}$, $\forall i$, and $\{v_{i,k}\}$ are i.i.d. Gaussian with zero mean and variance $R_i > 0$, $i = 1, \ldots, M$. The processes $\{v_{i,k}\}$ and $\{w_k\}$ are assumed to be mutually independent, with (A, C) detectable and $(A, Q^{1/2})$ stabilizable, where $C \triangleq \mathrm{col}(C_1, \ldots, C_M)$.

We assume that the measurements $y_{i,k}$ have undergone source coding and can be grouped into packets of b bits, with each packet short enough to be transmitted within one time step. In particular, the uniform quantizer of [12] will be used here. Under the additive noise model for quantization (which in general is quite accurate for bit rates as low as three bits per sample [13]), the quantized value of $y_{i,k}$ can be written as

$$y_{i,k}^q = y_{i,k} + q_{i,k},$$

where the quantization noise $q_{i,k}$ has variance $\delta_b \mathbb{E}[y_{i,k}^2]$, with

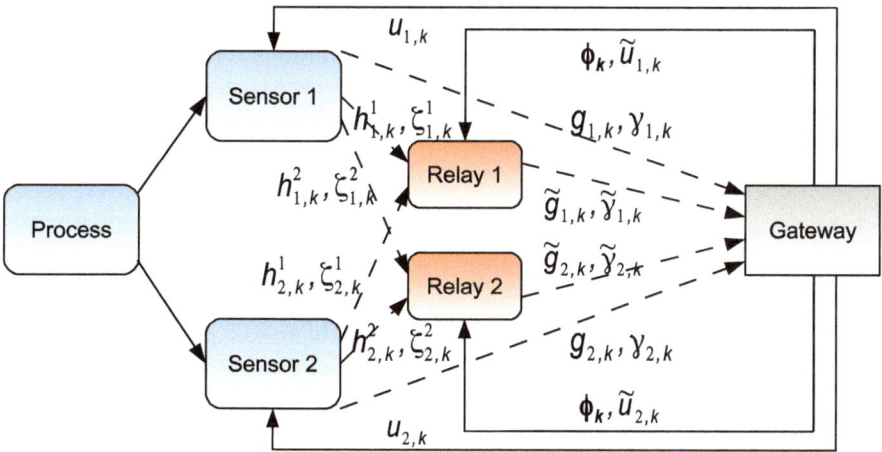

Fig. 5.1 System model for the case of two sensors and two relays

$$\delta_b = \frac{4b \ln 2}{3 \times 2^{2b}}$$

when using the uniform quantizer of [12]. The measurements $y_{i,k}^q$ are transmitted over parallel channels to a gateway, which will perform the remote state estimation. Let $\gamma_{i,k}$, for $i = 1, \ldots, M$, be random variables such that $\gamma_{i,k} = 1$ if $y_{i,k}^q$ is successfully transmitted to the gateway by sensor i, and $\gamma_{i,k} = 0$ otherwise.

Furthermore, there exist L intermediate relay nodes that can be used to aid the transmission of the sensor measurements to the gateway. Such situations can for instance occur in mesh networks, where nodes close to the process will make measurements of the process, while the other nodes do not make measurements, but can be used to relay the sensor measurements to the gateway [7]. A diagram of the system model for the case of $M = 2$ sensors and $L = 2$ relays is shown in Fig. 5.1. Each relay can listen (with possible dropouts) to a subset of the sensor transmissions. Denote $\mathbb{I} = \{1, \ldots, M\}$ as the set of all sensors, and $\mathbb{I}_l \subseteq \mathbb{I}$ as the set of sensors which relay l can listen to. In general, the sets $\mathbb{I}_l, l = 1, \ldots, L$ will not necessarily be disjoint, with possibly multiple relays listening to a given sensor. For $i \in \mathbb{I}_l$, let $\zeta_{i,k}^l$ be a random variable such that $\zeta_{i,k}^l = 1$ if the transmission at time k of sensor i is received by relay l, and $\zeta_{i,k}^l = 0$ otherwise. The relays can perform some simple local processing before transmitting over parallel channels to the gateway. Let $\tilde{\gamma}_{l,k}$ for $l = 1, \ldots, L$ be random variables such that $\tilde{\gamma}_{l,k} = 1$ if transmission at time k from relay l to the gateway is successful, and $\tilde{\gamma}_{l,k} = 0$ otherwise.

In this section we will consider a few simple operations that the relay can perform.[1] A relay can either (a) listen to one of the sensors' transmissions, say sensor i, and forward $y_{i,k}^q$ if it is successfully received by the relay, or (b) listen to a number of the sensors' transmissions, say sensors $i_1, i_2, \ldots, i_l$, and send $y_{i_1,k}^q \oplus y_{i_2,k}^q \oplus \cdots \oplus y_{i_l,k}^q$ if $y_{i_1,k}^q, y_{i_2,k}^q, \ldots, y_{i_l,k}^q$ have all been successfully received by the relay, where $\oplus$ is the XOR operation. The XOR operation is commonly used in network coding [1, 2]. For instance, if the gateway receives both $y_{i,k}^q$ and $y_{i,k}^q \oplus y_{j,k}^q$, then the gateway can recover $y_{j,k}^q$ using $y_{j,k}^q = y_{i,k}^q \oplus \left(y_{i,k}^q \oplus y_{j,k}^q \right)$. In general, given the transmissions received at the gateway, the measurements which can be recovered can be determined using Gaussian elimination over $\mathbb{Z}_2$.[2] Determining which sensor/s each relay listens to, and which operation each relay uses, is one of the key questions to be addressed in Sect. 5.1.5. We define a *relay configuration*

$$\boldsymbol{\phi}_k = (\phi_{1,k}, \ldots, \phi_{L,k})$$

at time k as the set of operations $\phi_{l,k}$ that each relay uses at time k. The set of all possible relay configurations will be denoted by $\boldsymbol{\Phi}$.

The communication channels will be modelled as time-varying fading channels. We let $g_{i,k}, i = 1, \ldots, M$ be the channel gains at time k from sensor i to the gateway, $\tilde{g}_{l,k}, l = 1, \ldots, L$ the channel gains from relay l to the gateway, and $h_{i,k}^l, i \in \mathbb{I}_l, l = 1, \ldots, M$ the channel gains from sensor i to relay l. We use the block fading model [14] and assume that $\{g_{i,k}\}, \{\tilde{g}_{l,k}\}, \{h_{i,k}^l\}$ vary over time k in an i.i.d. manner, with the processes being mutually independent. Denote the transmit powers at time k of the sensors and relays by $u_{i,k}, i = 1, \ldots, M$ and $\tilde{u}_{l,k}, l = 1, \ldots, L$, respectively. Following the model of Chap. 2, the packet reception probabilities will depend on both the channel gains and transmit powers as follows: We have $\lambda_{i,k} \triangleq \mathbb{P}(\gamma_{i,k} = 1 | g_{i,k}, u_{i,k}) = f(g_{i,k}u_{i,k})$ as the time-varying packet reception probabilities from sensor i to the gateway, $\tilde{\lambda}_{l,k} \triangleq \mathbb{P}(\tilde{\gamma}_{l,k} = 1 | \tilde{g}_{l,k}, \tilde{u}_{l,k}) = f(\tilde{g}_{l,k}\tilde{u}_{l,k})$ the probabilities from relay l to the gateway, and $\rho_{i,k}^l \triangleq \mathbb{P}(\zeta_{i,k}^l = 1 | h_{i,k}^l, u_{i,k}) = f(h_{i,k}^l u_{i,k})$ the probabilities from sensor i to relay l. Here

$$f(.) : [0, \infty) \to [0, 1] \tag{5.3}$$

is a continuous monotonically increasing function whose form depends on the particular digital modulation and coding scheme being used [15]. For example, in the case of uncoded binary phase shift keying (BPSK) transmission with b bits per packet,

---

[1] We assume limited computational power at the relays, thus only simple operations at a bit level are considered. If, however, additional computational capability is available, then other possibilities include the use of more involved network coding schemes [2] or the computation of local state estimates at the relays.

[2] While our use of the XOR operation is similar to network coding, our objectives are not exactly the same. In network coding transmissions are often regarded as "successful" only if all packets arrive (eventually) at their intended destinations, whereas in our problem even if some packets are not received one will still perform state estimation using the available measurements.

Table 5.1 Notation for different types of links

	Channel gain	Packet reception random variable	Packet reception probability
Sensor i to gateway	$g_{i,k}$	$\gamma_{i,k}$	$\lambda_{i,k}$
Relay l to gateway	$\tilde{g}_{i,k}$	$\tilde{\gamma}_{i,k}$	$\tilde{\lambda}_{i,k}$
Sensor i to relay l	$h^l_{i,k}$	$\zeta^l_{i,k}$	$\rho^l_{i,k}$

$f(.)$ would take the form

$$f(gu) = \left(\int_{-\infty}^{\sqrt{gu}} \frac{1}{\sqrt{2\pi}} e^{-t^2/2} dt \right)^b, \tag{5.4}$$

where we assume a packet is successfully received if and only if all b bits are successfully received. However, if there is channel coding and/or different digital modulation schemes, $f(.)$ will in general take on different forms [16]. In Table 5.1 we summarize the notation for the channel gains, packet reception random variables and packet reception probabilities for the different types of links.

In addition to the links carrying information from the sensors to the gateway, there are feedback links from the gateway to the sensors and relays which can be used to communicate the relay configuration ϕ_k and power levels $u_{i,k}$ and $\tilde{u}_{l,k}$ to be used, see Sects. 5.1.5 and 5.1.6. In this chapter we will assume that transmissions can occur over a much faster timescale than the process (5.1). Thus, delays experienced by the measurements in passing through intermediate relay nodes will be ignored. For instance, in the industrial wireless sensor networks standard WirelessHART [17], transmissions between nodes would typically take around 10 ms, whereas for many estimation and control applications the process time constant might be 1 s or more.

5.1.3 Kalman Filter with Packet Drops and Relays

Let $\theta_{i,k}, i = 1, \ldots, M$ be random variables such that $\theta_{i,k} = 1$ if $y^q_{i,k}$ can be reconstructed at the gateway, and $\theta_{i,k} = 0$ otherwise. Note that in general $\theta_{i,k}$ and $\theta_{j,k}, i \neq j$ are not independent even when the transmission dropouts are i.i.d. Values of $\theta_{i,k}$ for different relay configurations and combinations of $\gamma_{i,k}, \tilde{\gamma}_{l,k}, \zeta^l_{i,k}$ can be written in Boolean algebra form. For example, in Table 5.2 we give the Boolean expressions for $\theta_{1,k}$ and $\theta_{2,k}$ in the case of two sensors and one relay, where we use the notation $\wedge$ to denote logical 'and' and $\vee$ to denote logical 'or'. Now define

Table 5.2 Boolean expressions for $\theta_{1,k}$ and $\theta_{2,k}$, for two sensors and one relay, under three types of operations

ϕ_k	Forward $y_{1,k}^q$	Forward $y_{2,k}^q$	Send $y_{1,k}^q \oplus y_{2,k}^q$
$\theta_{1,k}$	$\gamma_{1,k} \vee (\bar{\gamma}_{1,k} \wedge \zeta_{1,k}^1)$	$\gamma_{1,k}$	$\gamma_{1,k} \vee (\bar{\gamma}_{1,k} \wedge \gamma_{2,k} \wedge \zeta_{1,k}^1 \wedge \zeta_{2,k}^1)$
$\theta_{2,k}$	$\gamma_{2,k}$	$\gamma_{2,k} \vee (\bar{\gamma}_{1,k} \wedge \zeta_{2,k}^1)$	$\gamma_{2,k} \vee (\bar{\gamma}_{1,k} \wedge \gamma_{1,k} \wedge \zeta_{1,k}^1 \wedge \zeta_{2,k}^1)$

$$\check{C}_k \triangleq \mathrm{col}(\theta_{1,k}C_1, \ldots, \theta_{M,k}C_M), \quad \check{y}_k \triangleq \mathrm{col}(\theta_{1,k}y_{1,k}^q, \ldots, \theta_{M,k}y_{M,k}^q),$$
$$\hat{x}_{k+1|k} \triangleq \mathbb{E}[x_{k+1}|\check{y}_0, \ldots, \check{y}_k, \check{C}_0, \ldots, \check{C}_k], \tag{5.5}$$
$$P_{k+1|k} \triangleq \mathbb{E}[(x_{k+1} - \hat{x}_{k+1|k})(x_{k+1} - \hat{x}_{k+1|k})^T|\check{y}_0, \ldots, \check{y}_k, \check{C}_0, \ldots, \check{C}_k].$$

The associated Kalman filter equations which are run at the gateway can be written as (see e.g. [5])

$$\hat{x}_{k+1|k} = A\hat{x}_{k|k-1} + K_k(\check{y}_k - \check{C}_k\hat{x}_{k|k-1})$$
$$P_{k+1|k} = AP_{k|k-1}A^T + Q - K_k\check{C}_kP_{k|k-1}A^T, \tag{5.6}$$

where $K_k = AP_{k|k-1}\check{C}_k^T(\check{C}_kP_{k|k-1}\check{C}_k^T + \check{R}_k)^{-1}$, with $\check{R}_k = \mathrm{diag}(\check{R}_{1,k}, \ldots, \check{R}_{M,k}) \triangleq \mathrm{diag}(R_1 + \delta_b\mathbb{E}[y_{1,k}^2], \ldots, R_M + \delta_b\mathbb{E}[y_{M,k}^2])$, similar to [6]. In the sequel, we will also call $P_k \triangleq P_{k|k-1}$.

5.1.4 Performance of the Kalman Filter with Relays

In Sect. 5.1.2 we have proposed that each relay can either listen to transmissions from one of the sensors which it then forwards to the gateway, or listen to a number of sensors and perform an XOR operation that is then sent to the gateway. We wish to investigate which operation each relay should use, and which sensors each relay should listen to, i.e. determining the relay configuration ϕ_k, in order to give the best performance for the Kalman filter. This section presents some preliminary results on the performance of the Kalman filter, with optimal relay configuration selection to be studied in Sect. 5.1.5. We consider the problem of optimal relay configuration selection in order to minimize the trace of the one step ahead expected error covariance $\mathbb{E}[P_{k+1}|P_k, \mathbf{g}_k, \phi_k]$, where

$$\mathbf{g}_k \triangleq \{g_{1,k}, \ldots, g_{M,k}, \tilde{g}_{1,k}, \ldots, \tilde{g}_{L,k}, h_{1,k}^1, \ldots, h_{M,k}^L\} \tag{5.7}$$

represents the channel gains at time k, which in turn will determine the packet reception probabilities $\lambda_{i,k}, \tilde{\lambda}_{l,k}, \rho_{i,k}^l, i = 1, \ldots, M, l = 1, \ldots, L$. In order to compute

$\mathbb{E}[P_{k+1}|P_k, \mathbf{g}_k, \boldsymbol{\phi}_k]$, we will further assume that full channel state information at the receiver is available, so that $\mathbf{g}_k$ is known at the gateway.[3]

Define

$$\mathfrak{F}_k(X) \triangleq AXA^T + Q - \mathbb{E}\left[AX\check{C}_k^T \left(\check{C}_k X \check{C}_k^T + \check{R}_k \right)^{-1} \check{C}_k X A^T \Big| \mathbf{g}_k, \boldsymbol{\phi}_k \right], \qquad (5.8)$$

where the expectation is with respect to $\theta_{1,k}, \ldots, \theta_{M,k}$ in the definition of $\check{C}_k$ in (5.5). Equivalently, we can write $\mathfrak{F}_k(X)$ as

$$\mathfrak{F}_k(X) = AXA^T + Q - \sum_{I \subseteq \mathbb{I}} \mathbb{E}\left[\prod_{i \in I} \Theta_{i,k} \prod_{j \notin I} (1 - \Theta_{j,k}) \Big| \mathbf{g}_k, \boldsymbol{\phi}_k \right] AX\bar{C}(I)^T \qquad (5.9)$$
$$\times \left(\bar{C}(I) X \bar{C}(I)^T + \bar{R}_k(I) \right)^{-1} \bar{C}(I) X A^T,$$

where $\bar{C}(I) = \mathrm{col}(\{C_i, i \in I\})$, $\bar{R}_k(I) = \mathrm{diag}(\{\check{R}_{i,k}, i \in I\})$, and $\Theta_{i,k}, i = 1, \ldots, M$ are random variables with the same distributions as $\theta_{i,k}$. The quantities $\mathbb{E}\left[\prod_{i \in I} \Theta_{i,k} \prod_{j \notin I} (1 - \Theta_{j,k}) \big| \mathbf{g}_k, \boldsymbol{\phi}_k \right]$ can be computed in terms of the packet reception probabilities $\lambda_{i,k}, \tilde{\lambda}_{l,k}, \rho_{i,k}^l, i = 1, \ldots, M, l = 1, \ldots, L$. A systematic procedure for doing this is as follows:

(1) Write out the Boolean expression

$$\bigwedge_{i \in I} \theta_{i,k} \bigwedge_{j \notin I} (\neg \theta_{j,k}), \qquad (5.10)$$

where each $\theta_{i,k}$ is written as a Boolean expression, $\neg \theta_{j,k}$ denotes the negation of the Boolean expression for $\theta_{j,k}$ and the notation $\bigwedge_{i \in I} \theta_{i,k} \triangleq \theta_{i_1,k} \wedge \cdots \wedge \theta_{i_n,k} \wedge \cdots$ for indices $i_n \in I$.

(2) Convert the Boolean expression (5.10) into the sum of products normal form [19]. Note that this can be done in a systematic way.

(3) $\mathbb{E}\left[\prod_{i \in I} \Theta_{i,k} \prod_{j \notin I} (1 - \Theta_{j,k}) | \mathbf{g}_k, \boldsymbol{\phi}_k \right]$ is then given by taking the sum of products normal form of (5.10), and replacing $\wedge$ with multiplication, $\vee$ with addition, $\gamma_{i,k}$ with $\lambda_{i,k}, \neg\gamma_{i,k}$ with $1 - \lambda_{i,k}, \tilde{\gamma}_{i,k}$ with $\tilde{\lambda}_{i,k}, \neg\tilde{\gamma}_{i,k}$ with $1 - \tilde{\lambda}_{i,k}, \zeta_{i,k}^l$ with $\rho_{i,k}^l$ and $\neg\zeta_{i,k}^l$ with $1 - \rho_{i,k}^l$.

By step (2) above, each term in the sum will correspond to a distinct entry of the truth table for $\theta_{1,k}, \ldots, \theta_{M,k}$, thus allowing $\mathbb{E}\left[\prod_{i \in I} \Theta_{i,k} \prod_{j \notin I} (1 - \Theta_{j,k}) | \mathbf{g}_k, \boldsymbol{\phi}_k \right]$ to be easily calculated.

We now give a result on how the packet reception probabilities affect the expected error covariance $\mathbb{E}[P_{k+1}|P_k, \mathbf{g}_k, \boldsymbol{\phi}_k] = \mathfrak{F}_k(P_k)$. First denote

---

[3] In practice, this can be achieved using channel estimation and prediction algorithms, see references in [6, 18].

$$\mathscr{L}_{i,k} \triangleq \{\lambda_{1,k}, \ldots, \lambda_{i,k}, \ldots, \lambda_{M,k}, \tilde{\lambda}_{1,k}, \ldots, \tilde{\lambda}_{L,k}, \rho_{1,k}^1, \ldots, \rho_{M,k}^L\}$$

$$\mathscr{U}_{i,k} \triangleq \{\lambda_{1,k}, \ldots, \mu_{i,k}, \ldots, \lambda_{M,k}, \tilde{\lambda}_{1,k}, \ldots, \tilde{\lambda}_{L,k}, \rho_{1,k}^1, \ldots, \rho_{M,k}^L\}$$

$$\tilde{\mathscr{L}}_{l,k} \triangleq \{\lambda_{1,k}, \ldots, \lambda_{M,k}, \tilde{\lambda}_{1,k}, \ldots, \tilde{\lambda}_{l,k}, \ldots, \tilde{\lambda}_{L,k}, \rho_{1,k}^1, \ldots, \rho_{M,k}^L\}$$

$$\tilde{\mathscr{U}}_{l,k} \triangleq \{\lambda_{1,k}, \ldots, \lambda_{M,k}, \tilde{\lambda}_{1,k}, \ldots, \tilde{\mu}_{l,k}, \ldots, \tilde{\lambda}_{L,k}, \rho_{1,k}^1, \ldots, \rho_{M,k}^L\}$$

$$\mathscr{R}_{i,k}^l \triangleq \{\lambda_{1,k}, \ldots, \lambda_{M,k}, \tilde{\lambda}_{1,k}, \ldots, \tilde{\lambda}_{L,k}, \rho_{1,k}^1, \ldots, \rho_{i,k}^l, \ldots, \rho_{M,k}^L\}$$

$$\mathscr{S}_{i,k}^l \triangleq \{\lambda_{1,k}, \ldots, \lambda_{M,k}, \tilde{\lambda}_{1,k}, \ldots, \tilde{\lambda}_{L,k}, \rho_{1,k}^1, \ldots, \sigma_{i,k}^l, \ldots, \rho_{M,k}^L\}.$$

Lemma 5.1 *Let $\mathfrak{F}_{\mathscr{X}_{i,k}}(.)$ be defined by $\mathfrak{F}_k(.)$ in (5.8) when the links have packet reception probabilities $\mathscr{X}_{i,k}$. Then, irrespective of which relay configuration is used, $\forall i = 1, \ldots, M, \forall l = 1, \ldots, L$, we have*

$$\lambda_{i,k} \leq \mu_{i,k} \Rightarrow \mathfrak{F}_{\mathscr{L}_{i,k}}(X) \geq \mathfrak{F}_{\mathscr{U}_{i,k}}(X)$$

$$\tilde{\lambda}_{l,k} \leq \tilde{\mu}_{l,k} \Rightarrow \mathfrak{F}_{\tilde{\mathscr{L}}_{l,k}}(X) \geq \mathfrak{F}_{\tilde{\mathscr{U}}_{l,k}}(X)$$

$$\rho_{i,k}^l \leq \sigma_{i,k}^l \Rightarrow \mathfrak{F}_{\mathscr{R}_{i,k}^l}(X) \geq \mathfrak{F}_{\mathscr{S}_{i,k}^l}(X).$$

Proof Consider the case $\lambda_{i,k} \leq \mu_{i,k}$. Recall that Bernoulli random variables can be generated from $U(0, 1)$ uniform random variables, by comparing the uniform random variable with the probability that the Bernoulli random variable is equal to one, i.e. $\gamma_{i,k} = 1$ when $u \leq \lambda_{i,k}$, and $\gamma_{i,k} = 0$ otherwise, where u is $U(0, 1)$. Let ω denote an outcome corresponding to N independent realizations of $U(0, 1)$ random variables, where N is equal to the total number of packet dropping links. For each ω, one can generate corresponding independent Bernoulli random variables $\gamma_{1,k}, \ldots, \gamma_{M,k}, \tilde{\gamma}_{1,k}, \ldots, \tilde{\gamma}_{L,k}, \zeta_{1,k}^1, \ldots, \zeta_{M,k}^L$. One can then construct the Bernoulli random variables $\theta_{1,k}, \ldots, \theta_{M,k}$, and hence $\check{C}_k$ as in (5.5).

Let $\check{C}_{\mathscr{L}_{i,k}}(\omega)$ be the matrix $\check{C}_k$ when using packet reception probabilities $\mathscr{L}_{i,k}$, and $\check{C}_{\mathscr{U}_{i,k}}(\omega)$ be the matrix $\check{C}_k$ when using packet reception probabilities $\mathscr{U}_{i,k}$. Now note that if $\theta_{j,k}(\omega) = 1$ using the packet reception probabilities $\mathscr{L}_{i,k}$, then we also have $\theta_{j,k}(\omega) = 1$ when using the packet reception probabilities $\mathscr{U}_{i,k}$, from the way in which $\theta_{j,k}(\omega)$ is constructed, and since an increase in the packet reception probability of any link cannot decrease the probability of reconstructing any of the sensor measurements. Hence

$$AX\check{C}_{\mathscr{L}_{i,k}}(\omega)^T \left(\check{C}_{\mathscr{L}_{i,k}}(\omega)X\check{C}_{\mathscr{L}_{i,k}}(\omega)^T + \check{R}_k\right)^{-1}\check{C}_{\mathscr{L}_{i,k}}(\omega)XA^T$$
$$\geq AX\check{C}_{\mathscr{U}_{i,k}}(\omega)^T \left(\check{C}_{\mathscr{U}_{i,k}}(\omega)X\check{C}_{\mathscr{U}_{i,k}}(\omega)^T + \check{R}_k\right)^{-1}\check{C}_{\mathscr{U}_{i,k}}(\omega)XA^T. \tag{5.11}$$

Since (5.11) holds for all ω, we have

$$\mathbb{E}\left[AX(\check{C}_{\mathscr{L}_{i,k}})^T\left(\check{C}_{\mathscr{L}_{i,k}}X(\check{C}_{\mathscr{L}_{i,k}})^T+\check{R}_k\right)^{-1}\check{C}_{\mathscr{L}_{i,k}}XA^T\Big|\mathbf{g}_k,\boldsymbol{\phi}_k\right]$$

$$\geq \mathbb{E}\left[AX(\check{C}_{\mathscr{U}_{i,k}})^T\left(\check{C}_{\mathscr{U}_{i,k}}X(\check{C}_{\mathscr{U}_{i,k}})^T+\check{R}_k\right)^{-1}\check{C}_{\mathscr{U}_{i,k}}XA^T\Big|\mathbf{g}_k,\boldsymbol{\phi}_k\right],$$

which shows that $\mathfrak{F}_{\mathscr{L}_{i,k}}(X) \geq \mathfrak{F}_{\mathscr{U}_{i,k}}(X)$. The other two cases can be proved in a similar manner. $\square$

5.1.5   Relay Configuration Selection

We now wish to address the question of determining which configurations for the relays will give the best Kalman filter performance. Suppose for now that the sensor transmit powers $u_{i,k}$, $i = 1, \ldots, M$ and relay transmit powers $\tilde{u}_{l,k}$, $l = 1, \ldots, L$ are given or fixed (The more difficult problem of jointly optimizing the relay configuration and transmission powers will be considered in Sect. 5.1.6.). We wish to choose at each time instant k, the relay configuration $\boldsymbol{\phi}_k^*$ that minimizes $\mathrm{tr}\mathbb{E}[P_{k+1}|P_k,\mathbf{g}_k,\boldsymbol{\phi}_k]$, i.e.

$$\boldsymbol{\phi}_k^* = \underset{\boldsymbol{\phi}_k(P_k,\mathbf{g}_k)\in\boldsymbol{\Phi}}{\mathrm{argmin}}\ \mathrm{tr}\mathbb{E}[P_{k+1}|P_k,\mathbf{g}_k,\boldsymbol{\phi}_k], \tag{5.12}$$

where $\mathbb{E}[P_{k+1}|P_k,\mathbf{g}_k,\boldsymbol{\phi}_k] = \mathfrak{F}_k(P_k)$, see (5.8).

Optimal Relay Configuration Selection

Problem (5.12) can, in principle, be solved by exhaustive search at the gateway. The optimal configuration can then be fed back to the relays. We will characterize the number of relay configurations that need to be checked at each time instant for exhaustive search.

Lemma 5.2 *Let $\mathbb{I}_l$ be the set of sensors that relay l can listen to, and let $M_l = |\mathbb{I}_l|$ denote the cardinality of $\mathbb{I}_l$. Suppose that there are no restrictions on how many relays listen to the same sensor. Then the number of possible relay configurations for $\boldsymbol{\phi}_k$ is*

$$|\boldsymbol{\Phi}| = \prod_{l=1}^{L}\left(2^{M_l} - 1\right). \tag{5.13}$$

Proof First fix a relay l, which can listen to M_l of the sensors. This relay can either forward any one of the sensor transmissions, or perform the XOR operation on two or more of the sensor transmissions it listens to, resulting in $M_l + \binom{M_l}{2} + \binom{M_l}{3} + \cdots + \binom{M_l}{M_l} = 2^{M_l} - 1$ possible operations. If there are no restrictions on multiple relays listening to the same sensor, then by the multiplication principle the number of relay configurations is $\prod_{l=1}^{L}\left(2^{M_l} - 1\right)$. $\square$

We thus see that the number of configurations that needs to be checked is, in the worst case (where each relay can listen to all sensors), exponential in the number of relays L and number of sensors M. However, in practice, due to geographical considerations, the number of sensors M_l that each sensor l listens to is often small, e.g. in [20, 21] it is assumed that $M_l \leq 3$.

Stability of Kalman Filtering with Optimal Relay Configuration Selection

We now wish to give a condition for stability of the Kalman filter with optimal relay configuration selection.

Definition 5.1 The Kalman filter is said to be *exponentially bounded* if there exist finite constants α and β, and an $r \in [0, 1)$, such that $\mathbb{E}[\operatorname{tr} P_k] \leq \alpha r^k + \beta$, $\forall k$.

Theorem 5.1 *Let $\{s_k\}$ be a stochastic process such that $s_k = 1$ if $\check{C}_k$ is full rank, and $s_k = 0$ otherwise. Suppose there exists a policy $\boldsymbol{\phi}^\sharp(\mathbf{g}_k)$ dependent only on $\mathbf{g}_k$, such that*

$$||A||^2 \mathbb{E}[\mathbb{P}(s_k = 0|\mathbf{g}_k, \boldsymbol{\phi}^\sharp(\mathbf{g}_k))] < 1, \tag{5.14}$$

where $||A||$ is the spectral norm of A. Then the Kalman filter using the optimal configurations obtained from problem (5.12) is exponentially bounded.

Proof Since the distribution of $\check{C}_k$ depends on P_k and $\mathbf{g}_k$, and $\mathbf{g}_k$ is independent in time and of P_k, we have $\mathbb{P}(\check{C}_k|P_k, P_{k-1}, \dots, P_0) = \mathbb{P}(\check{C}_k|P_k)$. Then by (5.6), the process $\{P_k\}$ is Markovian. Now define $V_k \triangleq \operatorname{tr} P_k$. In the relay configuration selection problem (5.12) we are minimizing $\mathbb{E}\{V_{k+1}|P_k, \mathbf{g}_k, \phi_k(P_k, \mathbf{g}_k)\}$. We thus have

$$
\begin{aligned}
\mathbb{E}\{V_{k+1}|P_k\} &= \mathbb{E}[\mathbb{E}\{V_{k+1}|P_k, \mathbf{g}_k, \boldsymbol{\phi}_k^*(P_k, \mathbf{g}_k)\}] \\
&\leq \mathbb{E}[\mathbb{E}\{V_{k+1}|P_k, \mathbf{g}_k, \boldsymbol{\phi}^\sharp(\mathbf{g}_k)\}] \\
&= \mathbb{E}\big[\mathbb{E}\{V_{k+1}|P_k, \mathbf{g}_k, \boldsymbol{\phi}^\sharp(\mathbf{g}_k), s_k = 1\}\mathbb{P}\{s_k = 1|P_k, \mathbf{g}_k, \boldsymbol{\phi}^\sharp(\mathbf{g}_k)\} \\
&\quad + \mathbb{E}\{V_{k+1}|P_k, \mathbf{g}_k, \boldsymbol{\phi}^\sharp(\mathbf{g}_k), s_k = 0\}\mathbb{P}\{s_k = 0|P_k, \mathbf{g}_k, \boldsymbol{\phi}^\sharp(\mathbf{g}_k)\}\big] \\
&\leq W + \big(||A||^2 V_k + \operatorname{tr} Q\big)\mathbb{E}[\mathbb{P}\{s_k = 0|\mathbf{g}_k, \boldsymbol{\phi}^\sharp(\mathbf{g}_k)\}],
\end{aligned}
$$

where the last inequality is shown using similar arguments to [6], and W is a positive constant. If $||A||^2\mathbb{E}[\mathbb{P}(s_k = 0|\mathbf{g}_k, \boldsymbol{\phi}^\sharp(\mathbf{g}_k))] < 1$ we may then use a stochastic Lyapunov function argument, similar to [6], to show that $\mathbb{E}\{V_k|P_0\} \leq \alpha r^k + \beta$, $\forall k$ for some $r \in [0, 1)$ and constants α and β, which establishes exponential boundedness of the Kalman filter. $\quad\square$

Theorem 5.1 thus provides a sufficient condition for Kalman filter stability dependent on the system matrix A and the distributions of the channel gains $\mathbf{g}_k$.

Example 5.1 Consider the case of two sensors and one relay, with $\check{C}_k$ being full rank only when both $\theta_{1,k} = \theta_{2,k} = 1$. Then

$$\mathbb{P}(\theta_{1,k} = 1, \theta_{2,k} = 1|\mathbf{g}_k, \boldsymbol{\phi}_k) = \mathbb{E}[\Theta_{1,k}\Theta_{2,k}|\mathbf{g}_k, \boldsymbol{\phi}_k] = \lambda_{1,k}\lambda_{2,k} + (1 - \lambda_{1,k})\lambda_{2,k}\tilde{\lambda}_{1,k}\rho_{1,k}^1.$$

Suppose we choose $\boldsymbol{\phi}^\sharp$ to be the suboptimal policy that always forwards $y_{1,k}^q$, and with the transmit powers $u_{1,k} = u_{2,k} = \tilde{u}_{1,k} = 1$. The condition (5.14) then becomes

$$\mathbb{E}[\mathbb{P}(s_k = 0 | \mathbf{g}_k, \boldsymbol{\phi}^\sharp(\mathbf{g}_k))] = \int \Big(1 - f(g_{1,k}) f(g_{2,k})$$

$$-(1 - f(g_{1,k})) f(g_{2,k}) f(\tilde{g}_{1,k}) f(h_{1,k}^1) \Big) d\mathbb{P}(\mathbf{g}_k) < \frac{1}{||A||^2},$$

which can be checked by numerically computing the integral for specific functions $f(.)$ in (5.3) and fading distributions $\mathbb{P}(\mathbf{g}_k)$. If condition (5.14) is satisfied for this suboptimal policy, then by Theorem 5.1 the Kalman filter using the optimal relay configurations will also be exponentially bounded.

Suboptimal Relay Configuration Selection

Lemma 5.2 has shown that the optimal way of choosing the relay configuration by checking each configuration is (at most) exponential in the number of relays L, which is computationally intensive when L is large. To reduce computational complexity, a suboptimal method for determining a relay configuration is to optimize the operation of each relay l independently of each other. A motivation for this method is that sometimes other relays may become unavailable; thus one should optimize the performance of each relay irrespective of what the other relays are doing. Specifically, consider subsets $I_l \subseteq \mathbb{I}_l$. Let $\bar{C}(I_l) = \text{col}(\{C_i, i \in I_l\})$, $\bar{R}_k(I_l) = \text{diag}(\{\check{R}_{i,k}, i \in I_l\})$, and

$$\mathfrak{F}_{l,k}(X) \triangleq AXA^T + Q - \sum_{I_l \subseteq \mathbb{I}_l} \mathbb{E}^l \left[\prod_{i \in I_l} \Theta_{i,k} \prod_{j \notin I_l} (1 - \Theta_{j,k}) | \mathbf{g}_k, \phi_{l,k} \right]$$

$$\times AX\bar{C}(I_l)^T \big(\bar{C}(I_l) X \bar{C}(I_l)^T + \bar{R}_k(I_l) \big)^{-1} \bar{C}(I_l) X A^T,$$

where the terms $\mathbb{E}^l \left[\prod_{i \in I_l} \Theta_{i,k} \prod_{j \notin I_l} (1 - \Theta_{j,k}) | \mathbf{g}_k, \phi_{l,k} \right]$ are computed assuming that relay l is the only relay available. One then computes $\mathfrak{F}_{l,k}(P_k)$ for each of the operations $\phi_{l,k}$ that relay l can perform, with the one that gives the smallest value of $\text{tr}(\mathfrak{F}_{l,k}(P_k))$ then chosen. This optimization can be carried out for each relay independently of the other relays. The number of configurations that need to be checked at each time step k is then $\sum_{l=1}^{L} (2^{M_l} - 1)$, which (compared to (5.13)) is no longer exponential in the number of relays L, and with M_l often being small in practice [20, 21].

5.1.6   Relay Configuration Selection and Power Control

In Sect. 5.1.5 the sensor and relay transmit powers were assumed to be fixed. However, similar to Chap. 2, the presence of time-varying fading channels will also allow for the use of power control techniques to further improve performance. In this section we present one possible formulation which optimizes the estimation performance subject to a sum of transmit powers constraint.

As in Sect. 5.1.5, we assume that full channel state information is available at the receiver, with $\mathbf{g}_k$ in (5.7) representing the set of all channel gains at time k. The transmit powers of the sensors and relays can then depend on both the instantaneous channel gains $\mathbf{g}_k$ and the error covariance P_k, with these transmit powers being computed at the gateway (which is assumed to have more computational resources than the sensors and relays) and fed back to the sensors and relays before transmission occurs. Denote $\mathbf{u}_k(\mathbf{g}_k, P_k) \triangleq \{u_{1,k}(\mathbf{g}_k, P_k), \ldots, u_{M,k}(\mathbf{g}_k, P_k), \tilde{u}_{1,k}(\mathbf{g}_k, P_k), \ldots, \tilde{u}_{L,k}(\mathbf{g}_k, P_k)\}$ as the set of all transmit powers at time k.

Optimal Power Control for a Given Relay Configuration

For a given relay configuration, we can pose the following power control problem:

$$\min_{\mathbf{u}_k(\mathbf{g}_k, P_k)} \mathrm{tr}\mathbb{E}[P_{k+1} | P_k, \mathbf{g}_k, \boldsymbol{\phi}_k]$$

$$\text{s.t.} \sum_{i=1}^{M} u_{i,k}(\mathbf{g}_k, P_k) + \sum_{l=1}^{L} \tilde{u}_{l,k}(\mathbf{g}_k, P_k) \leq u_{\mathrm{tot}}, \tag{5.15}$$

which minimizes the expected one step ahead error covariance subject to the sum power $\sum_{i=1}^{M} u_{i,k}(\mathbf{g}_k, P_k) + \sum_{l=1}^{L} \tilde{u}_{l,k}(\mathbf{g}_k, P_k)$ being less than a given bound u_{tot}. Due to the objective being a complicated nonlinear function of the transmit powers $\mathbf{u}_k$, the optimization problem (5.15) is in general non-convex and will need to be solved using numerical optimization algorithms.

Joint Relay Configuration Selection and Power Control

Problem (5.15) is for a given relay configuration $\boldsymbol{\phi}_k$. To optimally choose both the relay configuration and transmission powers, we can, in principle, solve $\prod_{l=1}^{L} (2^{M_l} - 1)$ instances of problem (5.15) at each time step (for each of the configurations, see Lemma 5.2), and choose the relay configuration that gives the smallest value for the objective function, which however is very computationally intensive.

A less computationally intensive suboptimal scheme is to first assume a simple power allocation (e.g. that the total power u_{tot} is equally divided between the sensors and relays), and use the suboptimal method of Sect. 5.1.5 to choose a relay configuration. For this chosen relay configuration, we then further optimize the transmission powers by solving the power control problem (5.15).

5.1.7 Numerical Studies

We first look at the performance of the optimal and suboptimal relay configuration selection methods of Sect. 5.1.5. We consider a situation with two sensors and two relays, where each of the relays can listen to both sensor transmissions, see Fig. 5.1. We consider the scalar case with $a = 0.95$, $q = 1$, $c_1 = c_2 = 1$, $r_1 = r_2 = 1$. For simplicity, we assume that the links from the sensors to the relays are perfect (i.e.

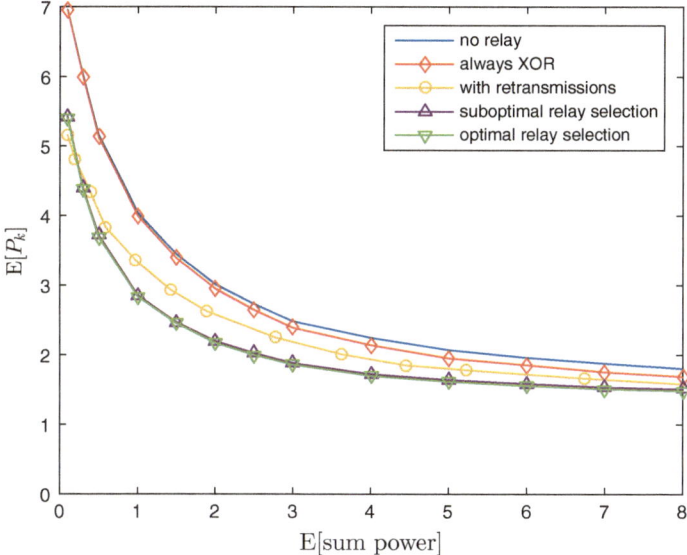

Fig. 5.2 Optimal and suboptimal relay configuration selections

have no dropouts), with the fading channels (from the sensors to gateway, and from the relays to gateway) being exponentially distributed with mean 1, which models the case of Rayleigh fading [15]. We assume that the digital communication uses BPSK transmission with $b = 6$ bits per packet, so that the function $f(.)$ in Sect. 5.1.2 has the form (5.4). We distribute the transmit powers equally between the sensors and relays. Figure 5.2 plots the average sum power and expected error covariance $\mathbb{E}[P_k]$ (obtained by time averaging $(x_k - \hat{x}_{k|k-1})(x_k - \hat{x}_{k|k-1})^T$ over 10,000 Monte Carlo iterations), for the optimal and suboptimal relay configuration selection methods. For comparison we also plot the performance for the cases of (1) no relay, (2) a scheme where the relay always performs the XOR operation as investigated in [6] and (3) a scheme where the gateway can ask for each lost transmission to be retransmitted once.[4] In each case, the expected error covariance decreases as the average power is increased. Since by (5.4) larger powers imply higher packet reception probabilities, this behaviour is in agreement with Lemma 5.1. We also see that the suboptimal method that optimizes each relay separately gives very close performance to the optimal method, and significantly outperforms the other schemes.

---

[4]Here we assume that additional transmit power (same as the power for a single transmission) is used for each retransmission, with a successfully retransmitted measurement (from time k) available to the Kalman filter at time $k + 1$, which now utilizes a buffer similar to [22].

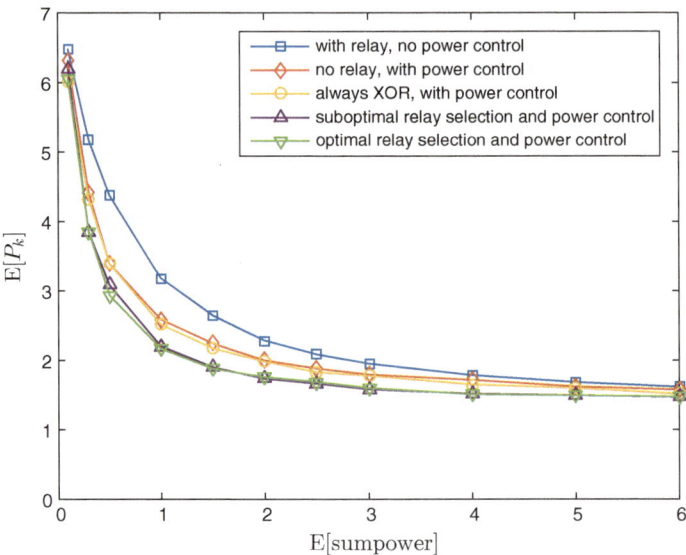

Fig. 5.3 Power control and relay configuration selections

We next consider the case of two sensors and one relay, with Rayleigh fading for each of the fading channels. We choose $g_{1,k}$, $g_{2,k}$ to have mean 1, while $\tilde{g}_{1,k}$, $h_{1,k}$, $h_{2,k}$ have mean 4. This models the case where power decays in free space as $1/d^2$, with d being the distance from the transmitter [15], and where the relay is located approximately halfway between the sensors and gateway. In Fig. 5.3 we plot $\mathbb{E}[P_k]$ (obtained by time averaging $(x_k - \hat{x}_{k|k-1})(x_k - \hat{x}_{k|k-1})^T$ over 10,000 Monte Carlo iterations) for different sum powers, obtained by solving problem (5.15) using the fmincon routine in MATLAB® for each relay configuration and selecting the best one. We also plot the performance of the suboptimal scheme where a relay config-uration is first chosen (assuming equal power allocation) and then power control is performed, see Sect. 5.1.6. We compare this with the case where there is no power control, with the sensors and relay using the same transmit power at all times, but with the best relay configuration chosen at each time step. Additionally, we plot the case where the relay always performs the XOR operation, and the case without a relay but with power control. We see that using power control gives significant performance benefits, with the best performance achieved when one optimizes both the relay configuration and transmit powers. The suboptimal scheme where a relay configuration is first chosen by assuming equal power allocation, and then the pow-ers are optimized, performs very close to the optimal scheme. Comparing the plots where power control is used, we see that for a given expected error covariance the average power required is significantly less (at least 30–40%) when a relay is used.

5.2 Network Topology Reconfiguration for Remote State Estimation

5.2.1 Background

Estimation in wireless sensor networks using a variety of different architectures has been considered in the literature. The architecture in [23] consists of one sensor making measurements, which is then transmitted over a lossy network with arbitrary topology. The article [24] looks at decentralized Kalman filtering with packet drops and/or delays. The works in [18, 25] consider one-hop transmission (or a star topology) over packet dropping links, with [25] investigating various different fusion rules, and [18] studying the effect of power control on stability. Sensor network architectures with relays are studied in [6] and Sect. 5.1, adopting network coding as a way to improve performance. Kalman filtering over networks with tree structures include [26–28], with [26] studying a stochastic sensor scheduling problem, and [27] studying routing algorithms and topology reconfiguration but in the absence of packet drops. In [28] the individual links in the tree can be packet dropping, and the notion of a *network state* process is introduced, which models random time variations in the wireless environment, for example, due to moving machines and robots in a factory.

In [28] the network topology, i.e. which sensors communicate to each other and how packets are routed through the network, is assumed to be fixed even over different network states. The current section differs from [28] in that we consider the problem of determining the optimal network topology configuration to use in each network state. We further assume that network topology reconfigurations do not occur instantly, but may incur a (temporal) cost, in that changing from one configuration to another, unwanted links will need to be removed before new links can be established [29]. This leads to a transient time where some links may not be available and poor transitory performance. The aim is to optimize an expected error covariance measure over the possible network configurations, taking into account this transient state when switching between different configurations.

5.2.2 System Model

As before, the process is a discrete-time linear system of the form[5]

$$x(k + 1) = Ax(k) + w(k), \quad k \in \mathbb{N}_0 \triangleq \{0, 1, 2, \ldots\},$$

where $x(k) \in \mathbb{R}^n$, and $w(k)$ is Gaussian with zero mean and covariance matrix $Q > 0$. The process is observed by M sensors, with measurements

---

[5]We change notation slightly by writing, e.g. $x(k)$ instead of x_k. This is to avoid having multiple subscripts in quantities such as $P(k_l)$ later on.

Fig. 5.4 Sensor network
with nine nodes. The set of
active links represented by
arrows forms a tree, while
the *dotted lines* represent
inactive links

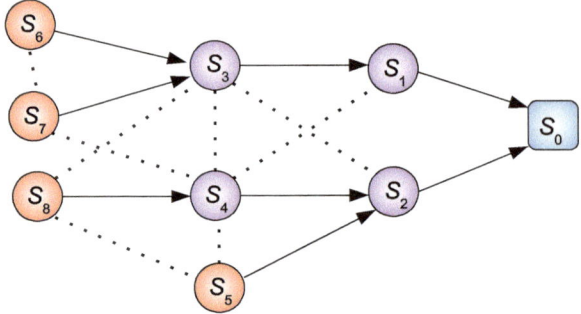

$$y_m(k) = C_m x(k) + v_m(k), \quad m \in \{1, \ldots, M\},$$

where $y_m(k) \in \mathbb{R}^{l_m}$, and $v_m(k)$ is Gaussian with zero mean and covariance matrix $R_m > 0$. We assume that $\{w\}$ and $\{v_m\}, m = 1, \ldots, M$ are i.i.d. over time and mutually independent. We make the assumption that (A, C) is detectable and $(A, Q^{1/2})$ is stabilizable, where $C \triangleq \mathrm{col}(C_1, \ldots, C_M)$. However, the individual (A, C_m) pairs are not required to be detectable.

Sensor Network Model

We consider the situation where some sensors and a gateway are connected to form a sensor network, which in general is assumed to have a mesh structure. Sensor measurements are to be transmitted, possibly via intermediate nodes, to the gateway, which performs the remote state estimation. The paths used by the sensors in transmitting to the gateway are usually computed using routing algorithms. We assume that the links which are utilized in the set of routes from the sensors to the gateway, which we denote as the set of *active links*, have a tree structure (i.e. has no cycles or parallel paths) with the gateway as the root node. This reduces redundancy in transmissions and energy usage, and avoids sensors having to listen to multiple transmissions.

The set of active links can be described using a directed graph with nodes/vertices $\{S_0, S_1, \ldots, S_M\}$, where the root node S_0 denotes the gateway, and $S_m, m = 1, \ldots, M$ denote the sensors. See Fig. 5.4 for an example with nine nodes (eight sensors and a gateway). Each sensor aggregates its own measurement to the received packets from incoming nodes and transmits the resulting packet to a single destination node. As in Sect. 5.1, we assume that transmissions can occur over a much faster timescale than the process, thus delays experienced in travelling through the network will be ignored in the sequel. We call the node that sensor S_m transmits to, the *parent* of S_m, denoted by $\mathrm{Par}(S_m)$. The outgoing link/edge from each of the nodes will be denoted as $\mathscr{E}_m = (S_m, \mathrm{Par}(S_m)), m = 1, \ldots, M$. For a given tree, there is a unique path from each node S_m to the gateway S_0, denoted by $\mathrm{Path}(S_m)$, with $\mathrm{Edges}(\mathrm{Path}(S_m))$ being the corresponding edges.

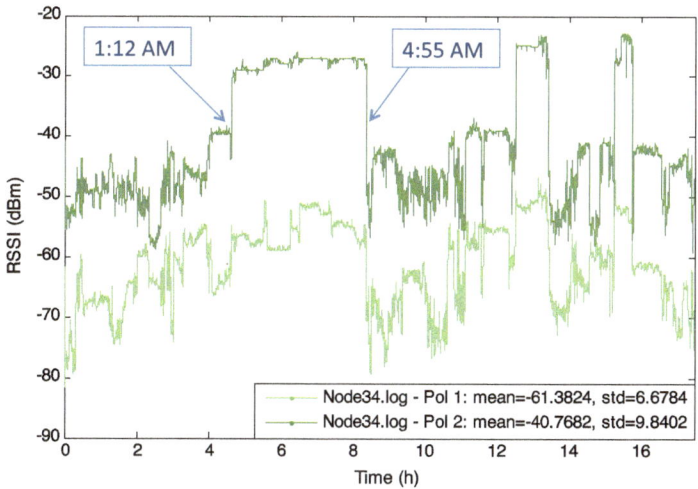

Fig. 5.5 Channel measurements taken at a paper mill

Wireless Channel Model

We model changes in the characteristics of the wireless environment by the notion of a randomly time-varying *network state* process $\mathcal{E}(k) \in \mathbb{B} \triangleq \{1, 2, \ldots, |\mathbb{B}|\}$. As motivation, consider Fig. 5.5, which plots some fading channel measurements acquired at a rolling mill in Iggesund, Sweden [30]. We see infrequent but substantial variations in the measured channel gains, due to mobile machinery and cranes in the ceiling blocking the line of sight between certain sensors, or changing the propagation pattern. Different network states can be used to represent the different positions (or similar groups of positions) that the machines are in.[6] We will assume that the network state process $\{\mathcal{E}\}$ is a discrete-time semi-Markov process [31], to model situations where network state transitions occur randomly, but not necessarily at every discrete-time instant k, see Fig. 5.6. The transition instants between network states are denoted by $\mathbb{K} \triangleq \{k_l\}$, with $k_0 = 0$, and $k_0 < k_1 < k_2 \cdots$ all integers. The *holding times*, or the amounts of time spent in a network state between transitions, are defined as $\Delta_l \triangleq k_{l+1} - k_l$. We will also refer to the period between successive network state transitions as a *holding period*. We assume that the holding times are bounded, thus $\Delta_l \leq \Delta_{\max}, \forall l$. Let $\mathbb{D} \triangleq \{1, 2, \ldots, \Delta_{\max}\}$. We have

$$\mathbb{P}\{\mathcal{E}(k_{l+1}) = j, \Delta_l = \delta | \mathcal{E}(k_0), \ldots, \mathcal{E}(k_{l-1}), \mathcal{E}(k_l) = i, k_0, \ldots, k_l\}$$
$$= \mathbb{P}\{\mathcal{E}(k_{l+1}) = j | \mathcal{E}(k_l) = i\}\mathbb{P}\{\Delta_l = \delta | \mathcal{E}(k_l) = i\} \tag{5.16}$$
$$= q_{ij}\psi_i(\delta), \quad \forall(k_l, \delta, i, j) \in \mathbb{K} \times \mathbb{D} \times \mathbb{B} \times \mathbb{B},$$

---

[6]In practice, network states $\mathcal{E}(k)$ can be estimated by either directly observing the positions of the machinery on the factory floor, or by using techniques to estimate variations in the radio environment [30].

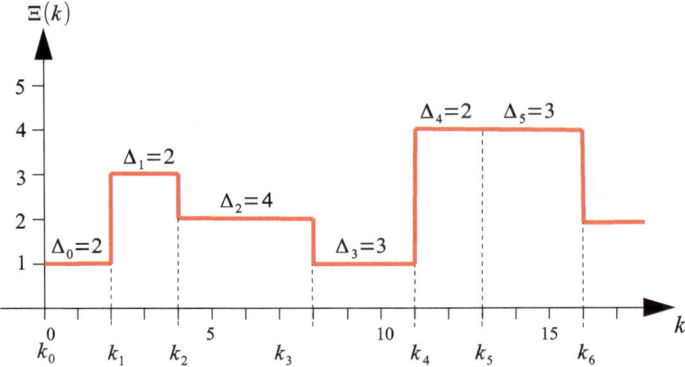

Fig. 5.6 Discrete-time semi-Markov process, see (5.16)

where in the second line we have made use of the fact that the Markov property holds at the transition instants (since the process is semi-Markov [31]), with

$$q_{ij} \triangleq \mathbb{P}\{\Xi(k_{l+1}) = j | \Xi(k_l) = i\}, \quad k_l, k_{l+1} \in \mathbb{K} \tag{5.17}$$

being the transition probabilities of the embedded Markov chain, and the fact that the conditional probabilities of the holding time

$$\psi_i(\delta) \triangleq \mathbb{P}\{\Delta_l = \delta | \Xi(k_l) = i\}, \quad k_l \in \mathbb{K} \tag{5.18}$$

depends only on the current state of the embedded Markov chain. We allow for virtual transitions [28] where $j = i$, see Fig. 5.6.

The *network configuration* $\pi(k)$ at time k fixes the transmission schedule that determines which nodes each sensor will receive from and forward to. The set of all possible network configurations is denoted by $\Pi = \{1, 2, \ldots, |\Pi|\}$, and the set of possible configurations when in network state j by $\Pi_j \subseteq \Pi$. We assume that the set of all possible network configurations has been precomputed and is known at the gateway. For instance, in each network state, one can compute a small number of reasonable configurations, using routing algorithms that optimize different objectives [32], which could also take into account possible link failures during operation. The set of all configurations in the different network states would then form our precomputed set of possible network configurations.

Define the random variables $\gamma_m(k)$, $m = 1, \ldots, M$ by

$$\gamma_m(k) = \begin{cases} 1, & \text{if transmission via link } \mathscr{E}_m \text{ at time } k \text{ is successful} \\ 0, & \text{otherwise}, \end{cases}$$

and the corresponding link success probabilities by

$$\phi_{m|(j,p)} \triangleq \mathbb{P}\{\gamma_m(k) = 1 | \Xi(k) = j, \pi(k) = p\}, \quad p \in \Pi_j.$$

We will assume that, *conditioned on a network state*, the dropouts $\{\gamma_m\}$ are i.i.d. Bernoulli processes, with $\{\gamma_m\}$ independent of $\{\gamma_n\}$ for $m \neq n$. Note that the packet reception probabilities can differ in different network states. Situations with i.i.d. and Markovian packet drops can also be regarded as special cases of this model, see [28] for details.

Kalman Filter at Gateway

Define the random variables $\theta_m(k), m = 1, \ldots, M$ by

$$\theta_m(k) = \begin{cases} 1, & \text{if transmission via Path}(S_m) \text{ at time } k \text{ is successful} \\ 0, & \text{otherwise,} \end{cases}$$

which determines whether the measurement of sensor m at time k is received by the gateway. Due to the fact that the set of active links forms a tree, we have

$$\theta_m(k) = \prod_{\mathscr{E}_i \in \text{Edges}(\text{Path}(S_m))} \gamma_i(k)$$

and, by independence,

$$\mathbb{P}\{\theta_m(k) = 1 | \varXi(k) = j, \pi(k) = p\} = \prod_{\mathscr{E}_i \in \text{Edges}(\text{Path}(S_m))} \phi_{i|(j,p)}.$$

Let $\theta(k) \triangleq \text{col}(\theta_1(k), \ldots, \theta_M(k))$, $y(k) \triangleq \text{col}(\theta_1(k)y_1(k), \ldots, \theta_M(k)y_M(k))$, $R \triangleq \text{diag}(R_1, \ldots, R_M)$, $C(k) \triangleq \text{col}(\theta_1(k)C_1, \ldots, \theta_M(k)C_M)$. The information set available at the gateway at time k is

$$\mathbb{I}(k) = \{\theta(0), \ldots, \theta(k), y(0), \ldots, y(k)\}.$$

The state estimates and estimation error covariances are defined as

$$\hat{x}(k|k-1) \triangleq \mathbb{E}\{x(k)|\mathbb{I}(k-1)\}.$$
$$P(k|k-1) \triangleq \mathbb{E}\{(x(k) - \hat{x}(k|k-1))(x(k) - \hat{x}(k|k-1))^T|\mathbb{I}(k-1)\}.$$

The Kalman filtering equations can then be written as (see [28])

$$\begin{aligned} \hat{x}(k+1|k) &= A\hat{x}(k|k-1) + K(k)(y(k) - C(k)\hat{x}(k|k-1)) \\ P(k+1|k) &= AP(k|k-1)A^T + Q - K(k)C(k)P(k|k-1)A^T, \end{aligned} \tag{5.19}$$

where $K(k) \triangleq AP(k|k-1)C(k)^T(C(k)P(k|k-1)C(k)^T + R)^{-1}$. In the sequel, we will also use the shorthand $P(k) \triangleq P(k|k-1)$.

5.2.3 Optimal Network Reconfiguration

As mentioned before, network states model random changes in the characteristics of the wireless environment. Due to these changes, see e.g. Fig. 5.5, the packet reception probabilities of existing links can change, and there could even be a complete (temporal) loss of connectivity in some links. The purpose of the present section is to illustrate how to compensate for changes in the wireless environment through network reconfiguration.

Reconfiguration Issues

In what follows, we will use a similar cost of reconfiguration as in [29], where in changing from one configuration to another, unwanted links will need to be removed before the establishment of new links. We will refer to this as a *transient state*. Thus there is a transient time or *reconfiguration time* $T_l \in \mathbb{N}_0$ at the lth state transition, where some links will not be available, resulting in poor transitory performance of the Kalman filter (see Sect. 5.2.5 for a specific example). Therefore, there is potentially a trade-off between choosing a configuration that gives good performance (after it is fully reconfigured) but requires many link changes, versus a configuration that has fewer link changes but poorer performance.

The reconfiguration time T_l is dependent on the underlying communication technology. For instance, in IEEE 802.11 the time needed to reroute a wireless network could be on the order of seconds, or even tens of seconds [33]. On the other hand, in WirelessHART which maintains multiple routes that can be switched at different time instances [34], it might be more appropriate to take $T_l = 0$. In this section T_l is taken to be random,[7] with a conditional probability distribution that could depend on the current network state $\Xi(k_l)$, the previous network configuration $\pi(k_{l-1})$, and the new network configuration chosen $\pi(k_l)$. We will assume that the reconfiguration times are bounded, i.e. $T_l \leq T_{\max}$, $\forall l$, for some finite $T_{\max}$.

Optimization Problem

At each transition instant $k_l \in \mathbb{K}$, we seek to find a network configuration

$$\pi(k_l) \triangleq \pi(P(k_l), \Xi(k_l), \pi(k_{l-1}))$$

which is to be held until the next transition instant $k_{l+1} \in \mathbb{K}$, and which minimizes an expected estimation error covariance performance measure over this holding period. The gateway decides on the new configuration based on knowledge of the current error covariance $P(k_l)$, the current network state $\Xi(k_l)$ and the old network configuration $\pi(k_{l-1})$, which is then communicated back to the sensors. For ease of exposition, we introduce the aggregated process

---

[7]Suppose the new configuration is to be communicated from the gateway back to the sensors (either using a broadcast or transmitted via intermediate nodes). Then, due to random packet losses, information about this new configuration may not get through reliably to all nodes at the same time but will need to be retransmitted, resulting in a random reconfiguration time T_l.

$$\mathscr{U}(k_l) \triangleq \left(P(k_l), \varXi(k_l), \pi(k_{l-1})\right), \quad k_l \in \mathbb{K}. \tag{5.20}$$

In terms of $\mathscr{U}(k_l)$, the new configuration $\pi^*(k_l) \in \Pi_j$ when $\varXi(k_l) = j$ is found via the following optimization:

$$\pi^*(k_l) = \mathrm{argmin}_{\pi(k_l) \in \Pi_j} \mathscr{V}(\mathscr{U}(k_l), \pi(k_l)), \tag{5.21}$$

where the cost function

$$\mathscr{V}(\mathscr{U}(k_l), \pi(k_l)) \triangleq \mathbb{E}\left\{ \sum_{d=1}^{\Delta_l} \mathrm{tr}P(k_l + d) \middle| \mathscr{U}(k_l), \pi(k_l) \right\}, \tag{5.22}$$

with the holding time Δ_l being random. The quantity $\mathscr{V}(\mathscr{U}(k_l), \pi(k_l))$ amounts to the sum of the trace of expected error covariances over the random holding time Δ_l, when the configuration $\pi(k_l)$ is used. In computations, it is useful to further rewrite (5.22) as

$$\begin{aligned}
\mathscr{V}(\mathscr{U}(k_l), \pi(k_l)) = \sum_{\delta=1}^{\Delta_{\max}} &\left[\sum_{t=0}^{T_{\max}} \mathbb{E}\left\{ \sum_{d=1}^{\delta} \mathrm{tr}P(k_l + d) \middle| \mathscr{U}(k_l), \pi(k_l), T_l = t \right\} \right.\\
&\left. \times \mathbb{P}\{T_l = t | \varXi(k_l) = j, \pi(k_{l-1}), \pi(k_l)\} \right] \mathbb{P}\{\Delta_l = \delta | \varXi(k_l) = j\}.
\end{aligned} \tag{5.23}$$

In (5.23), the expectations in the terms

$$\mathbb{E}\left\{ P(k_l + d) \mid \mathscr{U}(k_l), \pi(k_l), T_l = t \right\} \tag{5.24}$$

are taken over the packet loss processes (which affect the Kalman filter recursions (5.19)), while the summations over δ and t average over the random holding times and random reconfiguration times, respectively. Following the model of Sect. 5.2.2, the network state $\varXi(k_l)$ determines the distribution of the holding times (see (5.18)) and thereby the upper limit of the sum over d in (5.23); differences between the decision variable $\pi(k_l)$ and the previous configuration $\pi(k_{l-1})$ determine which links would be moved to a transient state. In particular, (5.24) is computed based on whether the network is still in the transient mode (if $d \leq T_l$) or has been fully reconfigured (if $d > T_l$), with the expectation taken over the discrete random variables $\{\theta(k_l), \ldots, \theta(k_l + d - 1)\}$.

Computational Aspects

In principle, problem (5.21) can be solved by checking the values of $\mathscr{V}(\mathscr{U}(k_l), \pi(k_l))$ for each of the different configurations $\pi(k_l) \in \Pi_j$. However, computation of the expectations in (5.24) involves considering the values of $P(k_l + d)$ for all possible combinations of $\{\theta(k_l), \ldots, \theta(k_l + d - 1)\}$, with the number of possibilities being $O(2^{Md})$ in general. In particular, computing $\mathbb{E}\{P(k_l + \Delta_{\max}) \mid \mathscr{U}(k_l), \pi(k_l), T_l\}$ will have a complexity of $O(2^{M \Delta_{\max}})$. Thus, for large holding times, which occur often in industrial settings, calculating the cost function (5.22) is computationally intensive. Section 5.2.4 proposes a suboptimal method, which minimizes an alternative cost function that can be computed with complexity $O(2^M)$.

Stochastic Stability Analysis

Before proceeding, we will present a criterion for estimator stability with network configurations provided by the optimal reconfiguration problem (5.21), by extending the methods developed in [28].

Definition 5.2 The Kalman filter is said to be uniformly bounded[8] if there exists a finite constant $B > 0$ such that $\mathbb{E}\{\mathrm{tr}\, P(k)\} \leq B, \forall k \in \mathbb{N}$.

First we have the following:

Lemma 5.3 *The process* $\{Z\}_{\mathbb{K}}$ *defined by*

$$Z(k_l) \triangleq (P(k_{l-1}+1), \ldots, P(k_l), \varXi(k_l), \pi(k_{l-1})), \quad k_l \in \mathbb{K}$$

is Markovian.

Proof Note that $\{\varXi\}_{\mathbb{K}}$ is Markovian, and $\pi(k_l)$ depends only on $(P(k_l), \varXi(k_l), \pi(k_l - 1))$. We also have

$$\mathbb{P}\{C(k_l)|P(k_l), \ldots, P(k_{l-1}+1), P(k_{l-1}), \ldots, \varXi(k_l), \varXi(k_{l-1}), \ldots, \pi(k_{l-1}), \pi(k_{l-2}), \ldots\}$$
$$= \mathbb{P}\{C(k_l)|P(k_l), \ldots, P(k_{l-1}+1), \varXi(k_l), \pi(k_{l-1})\}.$$

The result then follows from (5.19). □

Next, define the observability matrices $\mathscr{O}(k, k) = C(k)$,

$$\mathscr{O}(k+n, k) = \begin{bmatrix} C(k) \\ C(k+1)A \\ \vdots \\ C(k+n)A^n \end{bmatrix}, \quad n \in \mathbb{N}. \tag{5.25}$$

Consider the processes $\{\rho_d\}_{\mathbb{K}}, d = 1, \ldots, \Delta_l$, given by

$$\rho_d(k_l) = \begin{cases} 1 \text{ , if } \mathscr{O}(k_l + d - 1, k_l) \text{ is full rank} \\ 0 \text{ , otherwise.} \end{cases}$$

Taking into account the network state, network configurations and reconfiguration times, define

---

[8]Uniform boundedness can be easily shown to be equivalent to the notion of exponential boundedness adopted in Definition 5.1 and the preceding chapters.

$$\mu_d(j, p, p^-) \triangleq \mathbb{P}\{\rho_d(k_l) = 0 | \Xi(k_l) = j, \pi(k_l) = p, \pi(k_{l-1}) = p^-\}$$

$$= \sum_{t=0}^{T_{max}} \mathbb{P}\{\rho_d(k_l) = 0 | \Xi(k_l) = j, \pi(k_l) = p, \pi(k_{l-1}) = p^-, T_l = t\}$$

$$\times \mathbb{P}\{T_l = t | \Xi(k_l) = j, \pi(k_l) = p, \pi(k_{l-1}) = p^-\}.$$
(5.26)

Then we have the following theorem:

Theorem 5.2 *Suppose there exists a policy $\pi^\sharp(k_l) \triangleq \pi^\sharp(\Xi(k_l), \pi^\sharp(k_{l-1}))$, dependent only on the current network state $\Xi(k_l) = j$ and existing configuration $\pi^\sharp(k_{l-1}) = p^-$, such that*

$$\sum_{\delta=1}^{\Delta_{max}} \mu_\delta(j, \pi^\sharp(j, p^-), p^-)||A||^{2\delta}\psi_j(\delta) < 1, \ \forall j \in \mathbb{B}, \ \forall p^- \in \Pi,$$
(5.27)

where $||A||$ denotes the spectral norm of A, and $\psi_j(\delta)$ is as defined in (5.18). Then, under the optimal network reconfiguration method (5.21), the Kalman filter is uniformly bounded.

Proof See Appendix. □

Theorem 5.2 establishes a sufficient condition for estimator stability, see Sect. 5.2.5 for an example of how this condition can be verified numerically. Intuitively, condition (5.27) averages out non-full rank observation outcomes over the random holding times $\Delta_l = \delta$.

Remark 5.1 In the case of a single network state with i.i.d. packet drops, we have $\delta = 1$, and $\psi_j(\delta) = 1, \forall j$. Then $\mu_\delta(j, p, p^-)$ reduces to the probability that $C(k)$ is not full rank, and (5.27) becomes

$$\mathbb{P}\{C(k) \text{ is not full rank}\}||A||^2 < 1,$$

which is similar to the stability condition of [18]. Further reducing to a single sensor with C_1 being full rank, the probability of $C(k)$ not being full rank is the probability of dropping a packet, so (5.27) becomes

$$\mathbb{P}\{\gamma_1(k) = 0\}||A||^2 < 1,$$

which resembles the stability conditions of e.g. [35].

Remark 5.2 Theorem 5.2 differs from Theorem 2 of [28] in that the probabilities $\mu_d(j, p, p^-)$ also depends on the network configurations $\pi(k_{l-1})$ and $\pi(k_l)$, a concept which was not considered in [28]. In addition, $\mu_d(j, p, p^-)$ is defined to be a probability conditional on $\Xi(k_l)$ rather than $\Xi(k_{l-1})$, which is perhaps more natural since our chosen configurations depend on $\Xi(k_l)$ rather than $\Xi(k_{l-1})$.

Multiple Holding Periods

In (5.21) and (5.22) network reconfigurations are carried out by considering the sum of expected error covariances over one network state holding period (involving several time steps k). By looking further ahead over multiple holding periods, one can potentially achieve better performance. For the case of averaging over N holding periods, the new configuration $\pi(k_l) \in \Pi_j$ when $\Xi(k_l) = j$ is found via the following optimization:

$$
\begin{aligned}
\operatorname{argmin}_{\pi(k_l) \in \Pi_j} &\left[\mathbb{E}\left\{ \sum_{d_0=1}^{\Delta_l} \operatorname{tr} P(k_l + d_0) \middle| \mathscr{U}(k_l), \pi(k_l) \right\} \right. \\
&+ \min_{\pi(k_{l+1})} \mathbb{E}\left\{ \mathbb{E}\left\{ \sum_{d_1=1}^{\Delta_{l+1}} \operatorname{tr} P(k_{l+1} + d_1) \middle| \mathscr{U}(k_{l+1}), \pi(k_{l+1}) \right\} \middle| \mathscr{U}(k_l), \pi(k_l) \right\} + \cdots \\
&+ \left. \min_{\pi(k_{l+N-1})} \mathbb{E}\left\{ \mathbb{E}\left\{ \sum_{d_{N-1}=1}^{\Delta_{l+N-1}} \operatorname{tr} P(k_{l+N-1} + d_{N-1}) \middle| \mathscr{U}(k_{l+N-1}), \pi(k_{l+N-1}) \right\} \middle| \mathscr{U}(k_l), \pi(k_l) \right\} \right].
\end{aligned}
$$

(5.28)

Observe that in solving the multiple holding period optimal reconfiguration problems (5.28), we also obtain reconfiguration policies for $\pi(k_{l+1}), \ldots, \pi(k_{l+N-1})$. However, here we will adopt a moving horizon approach similar to [5], wherein the optimal $\pi^*(k_{l+1})$ will be obtained by solving problem (5.28) at the next transition instant $k_{l+1} \in \mathbb{K}$, the optimal $\pi^*(k_{l+2})$ is obtained by solving problem (5.28) at the transition instant k_{l+2} and so on. We note that optimization over N holding periods will require the computation of cost functions with an increased complexity of $O(2^{M\Delta_{\max}N})$.

5.2.4 Suboptimal Network Reconfiguration

To address the computational issues outlined in Sect. 5.2.3, in this section we study a suboptimal scheme which minimizes upper bounds to the expected error covariances, where these upper bounds can be computed recursively with lower complexity than the expected error covariance performance measure (5.22).

Optimization Problem

We adopt a suboptimal approach wherein, using $\mathscr{U}(k_l)$ defined as in (5.20), the new configuration $\bar{\pi}^*(k_l) \in \Pi_j$ is obtained via

$$
\bar{\pi}^*(k_l) = \operatorname{argmin}_{\pi(k_l) \in \Pi_j} \mathscr{W}(\mathscr{U}(k_l), \pi(k_l)), \tag{5.29}
$$

where

$$\mathscr{W}(\mathscr{U}(k_l), \pi(k_l)) \triangleq \sum_{\delta=1}^{\Delta_{\max}} \sum_{d=1}^{\delta} \mathrm{tr} Y(k_l + d) \mathbb{P}\{\Delta_l = \delta \mid \Xi(k_l) = j\}. \qquad (5.30)$$

The sequence $\{Y(k_l + 1), Y(k_l + 2), \ldots, Y(k_l + \Delta_{\max})\}$ is given by the following recursion:

$$\begin{aligned}
Y(k + 1) &= AY(k)A^T + Q \\
&\quad - \mathbb{E}\{AY(k)C(k)^T (C(k)Y(k)C(k)^T + R)^{-1} C(k)Y(k)A^T \mid \mathscr{U}(k_l), \pi(k_l)\} \\
&= AY(k)A^T + Q \\
&\quad - \sum_{t=0}^{T_{\max}} \mathbb{E}\{AY(k)C(k)^T (C(k)Y(k)C(k)^T + R)^{-1} C(k)Y(k)A^T \mid \mathscr{U}(k_l), \pi(k_l), T_l = t\} \\
&\quad \times \mathbb{P}\{T_l = t \mid \Xi(k_l) = j, \pi(k_l), \pi(k_{l-1})\},
\end{aligned}$$

$$(5.31)$$

with initial condition $Y(k_l) = P(k_l)$. The expectations

$$\mathbb{E}\{AY(k)C(k)^T (C(k)Y(k)C(k)^T + R)^{-1} C(k)Y(k)A^T \mid \mathscr{U}(k_l), \pi(k_l), T_l = t\},$$

for $k \in \{k_l, \ldots, k_l + \Delta_{\max} - 1\}$ in (5.31), are computed with respect to the random packet loss processes, taking into account whether the network is still in the transient mode ($k - k_l \le T_l$) or has been fully reconfigured ($k - k_l > T_l$), similar to the computation of (5.24). We have the following result:

Lemma 5.4 *The sequence $Y(k)$ is an upper bound to $\mathbb{E}\{P(k) \mid \mathscr{U}(k_l), \pi(k_l)\}$ for $k \ge k_l$.*

Proof Define

$$g_k(X) = AXA^T + Q - \mathbb{E}\{AXC(k)^T (C(k)XC(k)^T + R)^{-1} C(k)XA^T \mid \mathscr{U}(k_l), \pi(k_l)\}.$$

Lemma 5.4 is proved by using the fact that $g_k(.)$ is concave in X, and induction. The concavity of $g_k(.)$ can be shown by using similar techniques as in [35, 36]. The details are omitted for brevity. □

Thus, when the suboptimal method minimizes (5.30), what is minimized is not the expected error covariance performance measure (5.22), but by Lemma 5.4, an upper bound to (5.22).

Computational Aspects

Upper bounding sequences of the form (5.31) are much easier to compute than the expected error covariance when the holding times are large, since one now needs to consider $O(2^M)$ combinations of packet drops at each stage in (5.31), rather than $O(2^{M\Delta_{\max}})$ when computing the expected error covariance.[9] Furthermore, the bounds

---

[9]While still exponential in the number of sensors, for industrial settings with small subnetworks this is usually quite feasible.

often seem to be quite tight, see, e.g. [37].[10] In Sect. 5.2.6 we will see that in numerical simulations the configurations obtained using the suboptimal method are in many cases identical to the configurations obtained using the optimal method.

Stochastic Stability Analysis

We now give a stability condition for the suboptimal network reconfiguration method. First we have the following lemma:

Lemma 5.5 *The process* $\{\bar{Z}\}_{\mathbb{K}}$ *defined by*

$$\bar{Z}(k_l) \triangleq (Y(k_{l-1}+1), \ldots, Y(k_l), \Xi(k_l), \pi(k_{l-1})), \quad k_l \in \mathbb{K}$$

is Markovian.

Proof The proof follows from the fact that (1) $\{Y\}_{\mathbb{N}}$ is Markovian since $Y(k+1)$ depends only on $Y(k)$, (2) $\{\Xi\}_{\mathbb{K}}$ is Markovian, and (3) $\pi(k_l)$ depends only on $(Y(k_l), \Xi(k_l), \pi(k_l-1))$. $\square$

Now consider a process $\{s(k)\}$ defined by

$$s(k) = \begin{cases} 1 \text{ , if } C(k) \text{ is full rank} \\ 0 \text{ , otherwise.} \end{cases}$$

For $d = 1, \ldots, \Delta_l$, let

$$v_d(j, p, p^-) \triangleq \mathbb{P}\{s(k_l + d - 1) = 0 | \Xi(k_l) = j, \pi(k_l) = p, \pi(k_{l-1}) = p^-\}$$

$$= \sum_{t=0}^{T_{max}} \mathbb{P}\{s(k_l+d-1)=0|\Xi(k_l)=j, \pi(k_l)=p, \pi(k_{l-1})=p^-, T_l=t\}$$

$$\times \mathbb{P}\{T_l=t|\Xi(k_l)=j, \pi(k_l)=p, \pi(k_{l-1})=p^-\}.$$

We have the following theorem:

Theorem 5.3 *Suppose there exists a policy* $\pi^{\sharp}(k_l) \triangleq \pi^{\sharp}(\Xi(k_l), \pi^{\sharp}(k_{l-1}))$, *dependent only on* $\Xi(k_l) = j$ *and* $\pi^{\sharp}(k_{l-1}) = p^-$, *such that*

$$\sum_{\delta=1}^{\Delta_{max}} v_\delta(j, \pi^{\sharp}(j, p^-), p^-)||A||^{2\delta}\psi_j(\delta) < 1, \ \forall j \in \mathbb{B}, \ \forall p^- \in \Pi. \tag{5.32}$$

Then, under the suboptimal reconfiguration method (5.29), the Kalman filter is uniformly bounded (see Definition 5.2).

Proof See Appendix. $\square$

---

[10] Some tighter but more complicated bounds based on techniques in [38] can also be used.

Remark 5.3 Comparing Theorems 5.2 and 5.3, we see that the condition (5.32) in Theorem 5.3 involves probabilities of the matrices $C(k)$ not being full rank, which in general is larger than the probability of the observability matrices in (5.25) not being full rank. Thus the condition (5.32) in Theorem 5.3 is more stringent than condition (5.27) of Theorem 5.2.

Multiple Holding Periods

Similar to Sect. 5.2.3, for the case of averaging over N holding periods, the new configuration $\bar{\pi}^*(k_l) \in \Pi_j$ when $\Xi(k_l) = j$ is found via the following optimization:

$$
\operatorname{argmin}_{\pi(k_l)\in\Pi_j} \mathbb{E}\left\{ \sum_{d_0=1}^{\Delta_l} \operatorname{tr} Y_0(k_l+d_0) + \min_{\pi(k_{l+1})} \sum_{d_1=1}^{\Delta_{l+1}} \operatorname{tr} Y_1(k_{l+1}+d_1) + \cdots \right.
$$
$$
\left. + \min_{\pi(k_{l+N-1})} \sum_{d_{N-1}=1}^{\Delta_{l+N-1}} \operatorname{tr} Y_{N-1}(k_{l+N-1}+d_{N-1}) \right\}. \tag{5.33}
$$

The N sequences $\{Y_0(k_l+1), \ldots, Y_0(k_l+\Delta_{\max})\}, \ldots, \{Y_{N-1}(k_{l+N-1}+1), \ldots, Y_{N-1}(k_{l+N-1}+\Delta_{\max})\}$ in (5.33) are defined, for $n = 0, \ldots, N-1$, as follows:

$$
Y_n(k+1) = AY_n(k)A^T + Q - \sum_{t_n=0}^{T_{\max}} \mathbb{E}\{AY_n(k)C(k)^T(C(k)Y_n(k)C(k)^T+R)^{-1}
$$
$$
\times C(k)Y_n(k)A^T \mid \bar{\mathscr{U}}(k_{l+n}), \pi(k_{l+n}), T_{l+n} = t_n\}
$$
$$
\times \mathbb{P}\{T_{l+n} = t_n \mid \Xi(k_{l+n}), \pi(k_{l+n}), \pi(k_{l+n-1})\}, \tag{5.34}
$$

for $k \in \{k_{l+n}, \ldots, k_{l+n}+\Delta_{\max}-1\}$, with initial condition $Y_n(k_{l+n}) = Y_{n-1}(k_{l+n-1}+\Delta_{l+n-1}) = Y_{n-1}(k_{l+n})$. In (5.34), we have $\bar{\mathscr{U}}(k_l) \triangleq (P(k_l), \Xi(k_l), \pi(k_{l-1}))$, and $\bar{\mathscr{U}}(k_{l+n}) \triangleq (Y_n(k_{l+n}), \Xi(k_{l+n}), \pi(k_{l+n-1}))$ for $n > 0$. Note that in the suboptimal reconfiguration problem (5.33), the minimization over $\pi(k_{l+n})$ for $n > 0$ is computed based on $\bar{\mathscr{U}}(k_{l+n})$, rather than $\mathscr{U}(k_{l+n}) = (P(k_{l+n}), \Xi(k_{l+n}), \pi(k_{l+n-1}))$ as in the optimal method (5.28).

When looking over N holding periods, computation of the cost functions has a complexity of $O(2^{MN})$, which could be very intensive for large values of N. However, from numerical simulations, it appears that in many situations even the case $N = 1$ already provides most of the gains achieved by solving the N-period problem, see Sect. 5.2.6.

5.2.5 An Illustrative Example

Here we give an example to illustrate some of the concepts that have been introduced, in particular how to verify the stability condition (5.27) of Theorem 5.2. We will

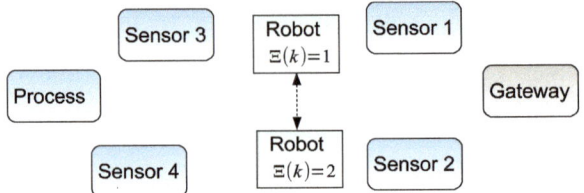

Fig. 5.7 Sensor network for example of Sect. 5.2.5

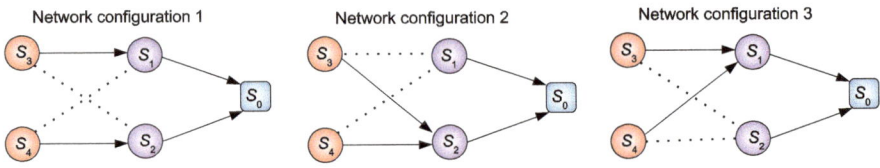

Fig. 5.8 Network configurations for example of Sect. 5.2.5

consider an example with four sensor nodes, see Fig. 5.7 for a diagram of the physical layout. The system has parameters

$$A = \begin{bmatrix} 1.1 & 0.2 \\ 0.2 & 0.8 \end{bmatrix}, \quad Q = \begin{bmatrix} 0.2 & 0 \\ 0 & 0.2 \end{bmatrix},$$

$C_1 = C_2 = C_3 = C_4 = \begin{bmatrix} 1 & 1 \end{bmatrix}$, $R_1 = R_2 = 20$, $R_3 = R_4 = 0.2$. The differences in the sensor measurement noise variances correspond to situations where either (i) the process is located much closer to sensors 3 and 4 than to sensors 1 and 2, or (ii) sensors 1 and 2 are located in a more hostile radio environment than sensors 3 and 4 [30]. However, sensors 1 and 2 have better connectivity to the gateway.

The set of all network configurations is shown in Fig. 5.8. There are two network states, with network configurations 1 and 2 possible when in network state 1 (so that $\Pi_1 = \{1, 2\}$), and network configurations 1 and 3 possible when in network state 2 (so that $\Pi_2 = \{1, 3\}$). The packet reception probabilities for the links in each of the network configurations are

$$
\begin{aligned}
\phi_{1|(1,1)} &= 0.5, \phi_{2|(1,1)} = 0.5, \phi_{3|(1,1)} = 0.1, \phi_{4|(1,1)} = 0.5 \\
\phi_{1|(1,2)} &= 0.5, \phi_{2|(1,2)} = 0.5, \phi_{3|(1,2)} = 0.8, \phi_{4|(1,2)} = 0.5 \\
\phi_{1|(2,1)} &= 0.5, \phi_{2|(2,1)} = 0.5, \phi_{3|(2,1)} = 0.5, \phi_{4|(2,1)} = 0.1 \\
\phi_{1|(2,3)} &= 0.5, \phi_{2|(2,3)} = 0.5, \phi_{3|(2,3)} = 0.5, \phi_{4|(2,3)} = 0.8.
\end{aligned}
\tag{5.35}
$$

Network state 1 corresponds to the case where there is a robot blocking the line of sight between sensor nodes 1 and 3, giving a packet reception probability of 0.1 for the direct link from sensor 3 to sensor 1 in network configuration 1, while in network configuration 2 sensor 3 will instead transmit to sensor 2 with a higher

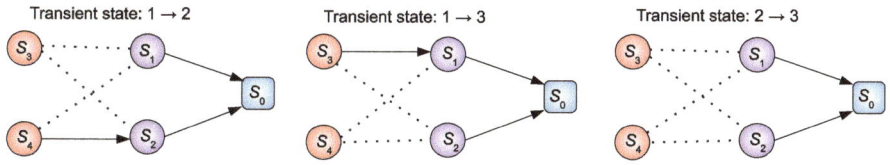

Fig. 5.9 Transient states when reconfiguring between two network configurations

packet reception probability of 0.8. Similarly, network state 2 will correspond to the case where the robot is now blocking the line of sight between sensors 2 and 4.

The transition probabilities for the embedded Markov chain $\{\Xi(k_l)\}$, $k_l \in \mathbb{K}$ are

$$\mathbb{P}\{\Xi(k_{l+1}) = 1 | \Xi(k_l) = 1\} = q_{11} = 0.5, \quad q_{12} = 0.5$$
$$\mathbb{P}\{\Xi(k_{l+1}) = 1 | \Xi(k_l) = 2\} = q_{21} = 0.5, \quad q_{22} = 0.5.$$

The reconfiguration times have the distribution:

$$\mathbb{P}\{T_l = 1 | \Xi(k_l), \pi(k_l), \pi(k_{l-1})\} = 0.8, \quad \mathbb{P}\{T_l = 2 | \Xi(k_l), \pi(k_l), \pi(k_{l-1})\} = 0.2,$$
$$(5.36)$$

$\forall (\Xi(k_l), \pi(k_l), \pi(k_{l-1}))$. The transient states in reconfiguring between different network configurations are shown in Fig. 5.9. For instance, in reconfiguring from network configuration 2 to configuration 3, the active links from sensor 3 to sensor 2, and from sensor 4 to sensor 2, will first need to be removed, leading to the transient state where sensors 3 and 4 do not have connectivity to the rest of the network for some time T_l. Similarly, reconfiguring from configuration 3 to configuration 2 will also lead to the same transient state.

We now illustrate how to verify the stability condition (5.27). We need to compute the terms $\mu_d(j, p, p^-)$, which, using (5.26), requires us to compute the probabilities

$$\mathbb{P}\{\rho_d(k_l) = 0 | \Xi(k_l) = j, \pi(k_l) = p, \pi(k_{l-1}) = p^-, T_l = t\}. \tag{5.37}$$

The observability matrices $\mathcal{O}(k_l + d - 1, k_l)$ are as in (5.25), where each $C(k) = \operatorname{col}(\theta_1(k)C_1, \ldots, \theta_M(k)C_M)$, $k = k_l, k_l + 1, \ldots, k_l + d - 1$. One can easily verify that if $\theta_{m_1}(k_1) = 1$ and $\theta_{m_2}(k_2) = 1$ for any $m_1, m_2 \in \{1, \ldots, M\}$, and any $k_1, k_2 \in \{k_l, k_l + 1, \ldots, k_l + d - 1\}$ with $k_1 \neq k_2$, then $\mathcal{O}(k_l + d - 1, k_l)$ has full rank. Thus $\mathcal{O}(k_l + d - 1, k_l)$ is not full rank when either:

(1) $\theta_m(k) = 0$, $\forall m \in \{1, \ldots, M\}$ and $\forall k \in \{k_l, k_l + 1, \ldots, k_l + d - 1\}$, or
(2) there exists a $k^* \in \{k_l, k_l + 1, \ldots, k_l + d - 1\}$ such that $\sum_{m=1}^{M} \theta_m(k^*) \geq 1$, and $\theta_m(k) = 0$, $\forall m \in \{1, \ldots, M\}$ and $k \neq k^*$.

First consider the instance $d = 4$, $\Xi(k_l) = 2$, $\pi(k_{l-1}) = 2$, $\pi(k_l) = 3$, $T_l = 1$. With these parameters, the network will be in the transient state $(2 \to 3)$ of Fig. 5.9 at time k_l, and be in network configuration 3 at times $k_l + 1, k_l + 2, k_l + 3$. Note that $\theta_m(k) = 0$, $\forall m$ when $\gamma_1(k) = 0$ and $\gamma_2(k) = 0$, both in the transient state and in network configuration 3. For case (1) above, note that for fixed k, the situation that $\theta_m(k) = 0, \forall m$ occurs with probability $(1 - \phi_{1|(2,3)})(1 - \phi_{2|(2,3)})$. Thus case (1) occurs with probability $\left[(1 - \phi_{1|(2,3)})(1 - \phi_{2|(2,3)})\right]^4$. For case (2) above, consider individually the four situations when $k^* = k_l, k_l + 1, k_l + 2, k_l + 3$. One can easily verify that each of these four situations occurs with probability

$$\left[1 - (1-\phi_{1|(2,3)})(1-\phi_{2|(2,3)})\right]\left[(1-\phi_{1|(2,3)})(1-\phi_{2|(2,3)})\right]^3,$$

and so case (2) occurs with probability

$$4\left[1 - (1-\phi_{1|(2,3)})(1-\phi_{2|(2,3)})\right]\left[(1-\phi_{1|(2,3)})(1-\phi_{2|(2,3)})\right]^3.$$

Hence

$$\mathbb{P}\{\rho_4(k_l) = 0 | \varXi(k_l) = 2, \pi(k_l) = 3, \pi(k_{l-1}) = 2, T_l = 1\} = \left[(1 - \phi_{1|(2,3)})(1 - \phi_{2|(2,3)})\right]^4$$
$$+ 4\left[1 - (1-\phi_{1|(2,3)})(1-\phi_{2|(2,3)})\right]\left[(1-\phi_{1|(2,3)})(1-\phi_{2|(2,3)})\right]^3.$$

Following the same arguments, it is not difficult to show that for other values of d, $\varXi(k_l) = j$, $\pi(k_{l-1}) = p^-$, $\pi(k_l) = p$, $T_l = t$, case (1) occurs with probability $\left[(1 - \phi_{1|(j,p)})(1 - \phi_{2|(j,p)})\right]^d$, case (2) occurs with probability

$$d\left[1 - (1-\phi_{1|(j,p)})(1 - \phi_{2|(j,p)})\right]\left[(1-\phi_{1|(j,p)})(1-\phi_{2|(j,p)})\right]^{d-1},$$

and hence

$$\mu_d(j, p, p^-) = \left[(1 - \phi_{1|(j,p)})(1 - \phi_{2|(j,p)})\right]^d$$
$$+ d\left[1 - (1-\phi_{1|(j,p)})(1 - \phi_{2|(j,p)})\right]\left[(1-\phi_{1|(j,p)})(1-\phi_{2|(j,p)})\right]^{d-1}. \tag{5.38}$$

Let the network state holding times have the following distribution:

$$\mathbb{P}\{\Delta_l = 1 | \varXi(k_l)\} = 0.1, \ \mathbb{P}\{\Delta_l = 2 | \varXi(k_l)\} = 0.1,$$
$$\mathbb{P}\{\Delta_l = 3 | \varXi(k_l)\} = 0.1, \ \mathbb{P}\{\Delta_l = 4 | \varXi(k_l)\} = 0.7, \ \forall \varXi(k_l). \tag{5.39}$$

Suppose $\pi^\sharp$ is the policy which uses network configuration 1 at all times. Then using (5.38) and (5.39), we find that for $j \in \{1, 2\}$,

$$\sum_{\delta=1}^{\Delta_{\max}} \mu_\delta(j, \pi^\sharp(j, p^-), p^-) \|A\|^{2\delta} \psi_j(\delta) = \sum_{\delta=1}^{\Delta_{\max}} \mu_\delta(j, 1, 1) \|A\|^{2\delta} \psi_j(\delta) = 0.4342 < 1.$$

Since we can find at least one policy satisfying condition (5.27) of Theorem 5.2, the Kalman filter with optimal reconfiguration will be uniformly bounded.

5.2.6 Numerical Studies

5.2.6.1 Comparison Between Optimal and Suboptimal Reconfiguration

We will use the same example as in Sect. 5.2.5, with the holding time distribution (5.39). The maximum holding time $\Delta_{\max} = 4$ is chosen to be small in order to allow for a comparison between the optimal and suboptimal reconfiguration methods of Sects. 5.2.3 and 5.2.4.

We first simulated a single realization of time length 10,000. The trace of the time averaged error covariance, $\mathbb{E}[\mathrm{tr}\,P(k)]$, when performing network reconfiguration is 1.65, whereas $\mathbb{E}[\mathrm{tr}\,P(k)]$ with no reconfiguration is 2.14, which amounts to a performance gain of about 30% for network reconfiguration. The network configurations obtained using both optimal and suboptimal methods behaved identically: Whenever the network state was equal to 1, the network changed to network configuration 2, while if the network state was equal to 2, the network changed to network configuration 3. However, different behaviours can be observed by modifying the packet reception probabilities. For instance, if in (5.35) both $\phi_{3|(1,1)}$ and $\phi_{4|(2,1)}$ are increased (so that the probability of packet reception in these two links for network configuration 1 is increased), then the network becomes less likely to reconfigure. From simulations, we found that for values of $\phi_{3|(1,1)}$ and $\phi_{4|(2,1)}$ greater than around 0.4, the network is always in network configuration 1, i.e. the network never reconfigures.

In Table 5.3 we give the values of $\mathbb{E}[\mathrm{tr}\,P(k)]$ under the optimal and suboptimal methods, for different values of $\phi_{3|(1,1)}$ and $\phi_{4|(2,1)}$, with $\phi_{3|(1,1)} = \phi_{4|(2,1)}$. Each $\mathbb{E}[\mathrm{tr}\,P(k)]$ entry is computed by taking the time average of Monte Carlo realizations of length 10,000. We also list the number of times when the optimal and suboptimal methods gave different network configurations. Only when $\phi_{3|(1,1)}$ and $\phi_{4|(2,1)}$ are around 0.3 did we observe significant differences (27 times in a realization of length 10,000) in the configurations obtained using the optimal and suboptimal methods, with the resulting performance being very similar. In terms of computational complexity, here $M = 4$, $\Delta_{\max} = 4$ and $T_{\max} = 2$. To compute the cost function for the optimal method requires consideration of approximately $(2^M + 2^{2M} + \cdots + 2^{\Delta_{\max}M}) \times T_{\max} = (2^4 + 2^8 + 2^{12} + 2^{16}) \times 2 = 139{,}808$ different terms. On the other hand, computing the cost function for the suboptimal method requires consideration of approximately $2^M \times T_{\max} \times \Delta_{\max} = 2^4 \times 2 \times 4 = 128$ different terms, substantially less than for the optimal method.

5.2.6.2 Comparison with Optimization over $N = 2$ Holding Periods

We now consider the case where the network state holding times have the following distribution:

$$\mathbb{P}\{\Delta_l = 11|\mathcal{E}(k_l)\} = \mathbb{P}\{\Delta_l = 12|\mathcal{E}(k_l)\} = \mathbb{P}\{\Delta_l = 13|\mathcal{E}(k_l)\} = \mathbb{P}\{\Delta_l = 14|\mathcal{E}(k_l)\} = \frac{1}{4},$$

Table 5.3 Comparison between optimal and suboptimal reconfiguration schemes

$\phi_{3|(1,1)} = \phi_{4|(2,1)}$	$\mathbb{E}[\operatorname{tr} P(k)]$ optimal	$\mathbb{E}[\operatorname{tr} P(k)]$ suboptimal	Differences in configurations b/w optimal and suboptimal
0.1	1.650	1.650	0
0.2	1.650	1.650	0
0.3	1.644	1.649	27
0.4	1.574	1.574	0
0.5	1.442	1.442	0
0.6	1.329	1.329	0
0.7	1.239	1.239	0
0.8	1.162	1.162	0
0.9	1.098	1.098	0

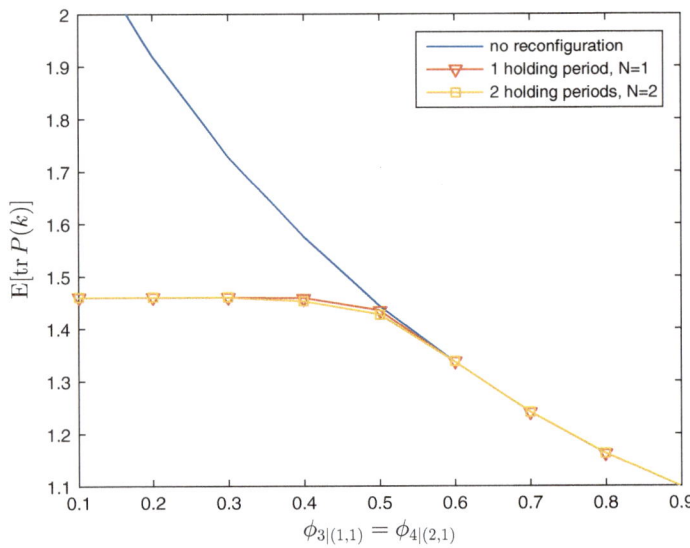

Fig. 5.10 $\mathbb{E}[\operatorname{tr} P(k)]$ for suboptimal network reconfiguration over one and two holding periods

$\forall \mathcal{E}(k_l)$, so that the minimum duration of a holding period is at least 11. Longer holding times are typically encountered in industrial environments, see e.g. Fig. 5.5. Due to the substantial increase in the computational complexity of solving the optimal reconfiguration problem for long holding times and/or the case of two holding periods, here we will only present results for the suboptimal methods of Sect. 5.2.4.

In Fig. 5.10 we plot $\mathbb{E}[\operatorname{tr} P(k)]$ when solving the suboptimal network reconfiguration problem over $N = 1$ or $N = 2$ holding periods, together with the case of no reconfiguration, for different values of $\phi_{3|(1,1)}$ and $\phi_{4|(2,1)}$, with $\phi_{3|(1,1)} = \phi_{4|(2,1)}$,

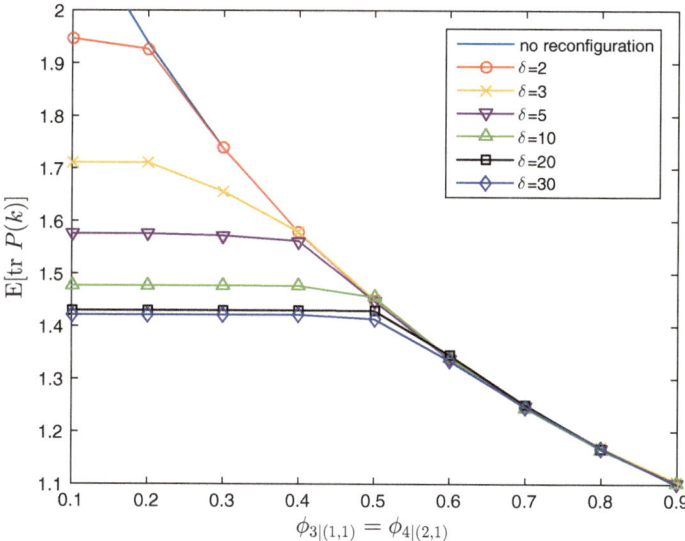

Fig. 5.11 $\mathbb{E}[\operatorname{tr} P(k)]$ for different holding times

where $\mathbb{E}[\operatorname{tr} P(k)]$ for each point on the graphs is obtained by taking the time average of Monte Carlo realizations of length 10,000. For small values of $\phi_{3|(1,1)}$ and $\phi_{4|(2,1)}$, the performance gains from reconfiguration are larger than in Sect. 5.2.6.1, due to the longer periods of time in which one can use a good network configuration before needing to reconfigure. For instance, when $\phi_{3|(1,1)} = \phi_{4|(2,1)} = 0.1$, $\mathbb{E}[\operatorname{tr} P(k)]$ is 1.46 with reconfigurations and 2.14 without reconfigurations, resulting in a performance gain of 47% for network reconfiguration, compared to 30% for the case examined in Sect. 5.2.6.1. We also see that the results are very similar when optimizing over both one or two holding periods. In fact, in our simulation results, only for values of $\phi_{3|(1,1)}$ and $\phi_{4|(2,1)}$ around 0.4–0.5 did we observe differences in the network configurations obtained, with the resulting performance differences being very small.

5.2.6.3 Performance Gains with Different Holding Times

We now consider the case where the network state holding times are fixed:

$$\mathbb{P}\{\Delta_l = \delta \mid \Xi(k_l) = 1\} = \mathbb{P}\{\Delta_l = \delta \mid \Xi(k_l) = 2\} = 1 \qquad (5.40)$$

for different values of δ. Figure 5.11 depicts $\mathbb{E}[\operatorname{tr} P(k)]$ in solving the suboptimal network reconfiguration problem over one holding period, for holding times of duration $\delta = 2, 3, 5, 10, 20, 30$, together with the case of no reconfiguration. We see that for larger δ, there is a greater performance gain by performing network reconfiguration. Additionally, there is a wider range of values of $\phi_{3|(1,1)}$ and $\phi_{4|(2,1)}$ where reconfiguration gives performance benefits. However, the relative performance gains diminish as δ increases, with little difference between the cases $\delta = 20$ and $\delta = 30$, due to the fact that reconfiguration times lose importance for large δ.

5.2.7   Conclusion

In this chapter we have first studied the use of relays for Kalman filtering with multiple sensors over packet dropping links, where the packet reception probabilities are governed by fading channel gains and sensor and relay transmit powers. By allowing relays to either forward one of the sensor's measurements or perform a network coding operation, we have considered the problem of determining the optimal relay configuration at each time step, together with a simpler suboptimal method. We have also studied the use of power control in addition to selecting the best relay configuration, to further improve performance. Numerical results have demonstrated that the use of relays can lead to significant power savings.

We then presented network topology reconfiguration methods for state estimation in sensor networks over time-varying wireless channels. The optimization of an expected error performance measure which takes into account the cost of reconfiguration has been studied. A less computationally intensive suboptimal method has been proposed, which in many cases gives identical results to the optimal method. In situations with long holding times, which are likely to be encountered in an industrial setting, numerical results suggest that significant performance gains can be achieved by network reconfiguration.

Notes: Section 5.1 is based on [39]. Section 5.2 is based on [40], which also describes some other low complexity reconfiguration approaches.

Appendix

Proof of Theorem 5.2

Consider a policy $\pi^\sharp$. Define the candidate stochastic Lyapunov function:

$$V_l \triangleq \operatorname{tr} P(k_l), \tag{5.41}$$

where $k_l \in \mathbb{K}$ are the switching times of the semi-Markov chain $\{\varXi\}$. We have

$$\mathbb{E}\{V_{l+1}|Z(k_l), \pi^\sharp(j, p^-)\} = \sum_{\delta=1}^{\Delta_{\max}} \mathbb{E}\{V_{l+1}|Z(k_l), \pi^\sharp(j, p^-), \Delta_l = \delta\}\psi_j(\delta).$$

$$\tag{5.42}$$

Noting that $k_{l+1} = k_l + \Delta_l$, we can write

$$\mathbb{E}\{V_{l+1}|Z(k_l), \pi^{\sharp}(j, p^-), \Delta_l = \delta\} = \mathbb{E}\{\mathrm{tr}\, P(k_l + \delta)|Z(k_l), \pi^{\sharp}(j, p^-)\}$$
$$= \mathbb{E}\{\mathrm{tr}\, P(k_l + \delta)|Z(k_l), \pi^{\sharp}(j, p^-), \rho_\delta(k_l) = 1\}\mathbb{P}\{\rho_\delta(k_l) = 1|Z(k_l), \pi^{\sharp}(j, p^-)\}$$
$$+ \mathbb{E}\{\mathrm{tr}\, P(k_l + \delta)|Z(k_l), \pi^{\sharp}(j, p^-), \rho_\delta(k_l) = 0\}\mathbb{P}\{\rho_\delta(k_l) = 0|Z(k_l), \pi^{\sharp}(j, p^-)\}.$$

For the case when $\rho_\delta(k_l) = 1$, by bounding the performance with that of a simple suboptimal predictor, we can show using similar arguments to [28] that

$$\mathbb{E}\{\mathrm{tr}\, P(k_l + \delta)|Z(k_l), \pi^{\sharp}(j, p^-), \rho_\delta(k_l) = 1\}\mathbb{P}\{\rho_\delta(k_l) = 1|Z(k_l), \pi^{\sharp}(j, p^-)\} \leq W_1^\delta$$

for some finite constant W_1^δ. For the case when $\rho_\delta(k_l) = 0$, the error covariance matrix $P(k_l + \delta)$ is bounded by the worst case where $\gamma_m(k) = 0, \forall (m, k) \in \{1, \ldots, M\} \times \{k_l, \ldots, k_l + \delta - 1\}$. Therefore,

$$\mathbb{E}\{\mathrm{tr}\, P(k_l + \delta)|Z(k_l), \pi^{\sharp}(j, p^-), \rho_\delta(k_l) = 0\}\mathbb{P}\{\rho_\delta(k_l) = 0|Z(k_l), \pi^{\sharp}(j, p^-)\}$$
$$\leq \mathrm{tr}\left(A^\delta P(k_l)(A^\delta)^T + A^{\delta-1}Q(A^{\delta-1})^T + \cdots + Q\right)\mu_\delta(j, \pi^{\sharp}(j, p^-), p^-)$$
$$\leq \mathrm{tr}\, P(k_l)||A||^{2\delta}\mu_\delta(j, \pi^{\sharp}(j, p^-), p^-) + W_2^\delta = V_l||A||^{2\delta}\mu_\delta(j, \pi^{\sharp}(j, p^-), p^-) + W_2^\delta$$
$$\tag{5.43}$$

for some finite constant W_2^δ. Then

$$\mathbb{E}\{V_{l+1}|Z(k_l), \pi^{\sharp}(j, p^-)\}$$
$$\leq \sum_{\delta=1}^{\Delta_{\max}} \mu_\delta(j, \pi^{\sharp}(j, p^-), p^-)||A||^{2\delta}\psi_j(\delta)V_l + \sum_{\delta=1}^{\Delta_{\max}} \left(W_1^\delta + W_2^\delta\right)\psi_j(\delta).$$

The second summation above is bounded. Thus, if

$$\sum_{\delta=1}^{\Delta_{\max}} \mu_\delta(j, \pi^{\sharp}(j, p^-), p^-)||A||^{2\delta}\psi_j(\delta) < 1$$

then, since $\{Z\}_{\mathbb{K}}$ is Markovian by Lemma 5.3, we can use Proposition 3.2 of [41] to show that, under the policy $\pi^{\sharp}$,

$$\mathbb{E}\{P(k_l)|Z(0), \pi^{\sharp}\} \leq \alpha_1 r^{k_l} + \beta_1, \quad \forall k_l \in \mathbb{K} \tag{5.44}$$

for some $r \in [0, 1)$ and finite constants α_1 and β_1. For the times in between transition instants, note that similar to (5.43), we can find finite constants α_2 and W_3 such that

$$\mathrm{tr}\, P(k_l + d) \leq ||A||^{2d}\mathrm{tr}\, P(k_l) + W_3 \leq \alpha_2 r^d \mathrm{tr}\, P(k_l) + W_3$$

holds for all $d \in \{1, \ldots, \Delta_l\}$. Then, using (5.44),

$$\mathbb{E}\{\mathrm{tr}\, P(k_l + d)|\pi^\sharp\} \leq \alpha_2\alpha_1 r^{k_l+d} + \alpha_2\beta_1 r^d + W_3 \leq \alpha r^{k_l+d} + \beta, \quad \forall d \in \{1, \ldots, \Delta_{\max}\}$$

for some finite constants α and β. Since $r < 1$, this implies that

$$\mathbb{E}[\mathrm{tr}\, P(k)|\pi^\sharp] \leq \alpha + \beta \triangleq B.$$

This establishes uniform boundedness at all times $k \in \mathbb{N}$ under policy $\pi^\sharp$ when condition (5.27) is satisfied.

Now, under the optimal reconfiguration policy π^*, we have

$$\mathbb{E}\{\mathrm{tr}\, P(k_l + 1) + \cdots + \mathrm{tr}\, P(k_l + \Delta_l)|Z(k_l), \pi^*\}$$
$$\leq \mathbb{E}\{\mathrm{tr}\, P(k_l + 1) + \cdots + \mathrm{tr}\, P(k_l + \Delta_l)|Z(k_l), \pi^\sharp\}$$

for all $Z(k_l)$, so that

$$\mathbb{E}\{\mathrm{tr}\, P(k_l + 1) + \cdots + \mathrm{tr}\, P(k_l + \Delta_l)|\pi^*\} \leq \mathbb{E}\{\mathrm{tr}\, P(k_l + 1) + \cdots + \mathrm{tr}\, P(k_l + \Delta_l)|\pi^\sharp\}$$
$$\leq \Delta_l B \leq \Delta_{\max} B.$$

Since error covariance matrices have nonnegative trace, we have for all $d \in \{1, \ldots, \Delta_l\}$,

$$\mathbb{E}\{\mathrm{tr}\, P(k_l + d)|\pi^*\} \leq \Delta_{\max} B \triangleq \tilde{B}.$$

This thus establishes uniform boundedness of the Kalman filter under the optimal policy π^*.

Proof of Theorem 5.3

First, for $k \in \{k_l, \ldots, k_l + \Delta_{\max} - 1\}$, the recursion (5.31) can be written as

$$\begin{aligned}
Y(k+1) = {} & \mathbb{E}\{AY(k)A^T + Q - AY(k)C(k)^T(C(k)Y(k)C(k)^T + R)^{-1} \\
& \times C(k)Y(k)A^T|\mathscr{U}(k_l), \pi(k_l), s(k) = 1\}\mathbb{P}\{s(k) = 1|\mathscr{U}(k_l), \pi(k_l)\} \\
& + \mathbb{E}\{AY(k)A^T + Q - AY(k)C(k)^T(C(k)Y(k)C(k)^T + R)^{-1} \\
& \times C(k)Y(k)A^T|\mathscr{U}(k_l), \pi(k_l), s(k) = 0\}\mathbb{P}\{s(k) = 0|\mathscr{U}(k_l), \pi(k_l)\},
\end{aligned}$$

from which one can derive the bounds

$$\mathrm{tr}Y(k_l + 1) \leq W_{1,1} + \mathrm{tr}\left(AY(k_l)A^T + Q\right)\mathbb{P}\{s(k_l) = 0|\mathcal{U}(k_l), \pi(k_l)\}$$
$$= W_{1,1} + \left(\mathrm{tr}Y(k_l)||A||^2 + W_{1,2}\right)\nu_1(\varXi(k_l), \pi(k_l), \pi(k_{l-1}))$$
$$\mathrm{tr}Y(k_l + 2) \leq W_{1,2} + \mathrm{tr}\left(AY(k_l + 1)A^T + Q\right)\mathbb{P}\{s(k_l + 1) = 0|\mathcal{U}(k_l), \pi(k_l)\}$$
$$\leq W_{1,2} + \left(\mathrm{tr}Y(k_l)||A||^4 + W_{2,2}\right)\nu_2(\varXi(k_l), \pi(k_l), \pi(k_{l-1}))$$
$$\vdots$$
$$\mathrm{tr}Y(k_l + d) \leq W_{1,d} + \left(\mathrm{tr}Y(k_l)||A||^{2d} + W_{2,d}\right)\nu_d(\varXi(k_l), \pi(k_l), \pi(k_{l-1}))$$

$$(5.45)$$

for some finite constants $W_{1,d}$ and $W_{2,d}$.

Consider a policy $\pi^\sharp$. Define

$$\bar{V}_l \triangleq \mathrm{tr}Y^\sharp(k_{l-1} + \Delta_{l-1}),$$

where $Y^\sharp$ denotes the recursion (5.31) under policy $\pi^\sharp$. We have

$$\mathbb{E}\{\bar{V}_{l+1}|\bar{Z}(k_l), \pi^\sharp(j, p^-)\} = \sum_{\delta=1}^{\Delta_{\max}} \mathbb{E}\{\bar{V}_{l+1}|\bar{Z}(k_l), \pi^\sharp(j, p^-), \Delta_l = \delta\}\psi_j(\delta)$$

and

$$\mathbb{E}\{\bar{V}_{l+1}|\bar{Z}(k_l), \pi^\sharp(j, p^-), \Delta_l = \delta\}$$
$$= \mathbb{E}\{\mathrm{tr}Y^\sharp(k_l + \delta)|\bar{Z}(k_l), \pi^\sharp(j, p^-), s(k_l + \delta - 1) = 1\}$$
$$\times \mathbb{P}\{s(k_l + \delta - 1) = 1|\bar{Z}(k_l), \pi^\sharp(j, p^-)\}$$
$$+ \mathbb{E}\{\mathrm{tr}Y^\sharp(k_l + \delta)|\bar{Z}(k_l), \pi^\sharp(j, p^-), s(k_l + \delta - 1) = 0\}$$
$$\times \mathbb{P}\{s(k_l + \delta - 1) = 0|\bar{Z}(k_l), \pi^\sharp(j, p^-)\}$$
$$\leq W_{1,\delta} + \left(\mathrm{tr}Y^\sharp(k_l)||A||^{2\delta} + W_{2,\delta}\right)\nu_\delta(j, \pi^\sharp(j, p^-), p^-)$$

for some finite constants $W_{1,\delta}$ and $W_{2,\delta}$, where the inequality comes from making use of the bounds (5.45). Using Lemma 5.5 and similar arguments as in the proof of Theorem 5.2, we can show that if

$$\sum_{\delta=1}^{\Delta_{\max}} \nu_\delta(j, \pi^\sharp(j, p^-), p^-)||A||^{2\delta}\psi_j(\delta) < 1$$

holds, then there exists finite constants B, $\tilde{B}$, such that $\mathrm{tr}Y^\sharp(k) \leq B$, $\forall k$ under policy $\pi^\sharp$, and $\mathrm{tr}Y(k) \leq \tilde{B}$, $\forall k$ under policy $\bar{\pi}^*$. Since $Y(k)$ upper bounds $\mathbb{E}\{P(k)\}$ by Lemma 5.4, this then implies $\mathbb{E}\{\mathrm{tr}P(k)\} \leq \tilde{B}$, $\forall k$, and hence the uniform boundedness of the Kalman filter with reconfiguration under the suboptimal policy $\bar{\pi}^*$.

References

1. S.-Y.R. Li, R.W. Yeung, N. Cai, Linear network coding. IEEE Trans. Inf. Theory **49**(2), 371–381 (2003)
2. T. Ho, D.S. Lun, *Network Coding: An Introduction* (Cambridge University Press, Cambridge, 2008)
3. Y. Hua, D.W. Bliss, S. Gazor, Y. Rong, Y. Sung (eds.), Special issue on theories and methods for advanced wireless relays. IEEE J. Sel. Areas Commun. **30**(8) (2012)
4. J.F. Kurose, K. Ross, *Computer Networking: A Top-Down Aproach*, 6th edn. (Pearson, Boston, 2012)
5. D.E. Quevedo, A. Ahlén, J. Østergaard, Energy efficient state estimation with wireless sensors through the use of predictive power control and coding. IEEE Trans. Signal Process. **58**(9), 4811–4823 (2010)
6. D.E. Quevedo, J. Østergaard, A. Ahlén, Power control and coding formulation for state estimation with wireless sensors. IEEE Trans. Control Syst. Technol. **22**(2), 413–427 (2014)
7. J.N. Laneman, D.N.C. Tse, G.W. Wornell, Cooperative diversity in wireless networks: efficient protocols and outage behavior. IEEE Trans. Inf. Theory **50**(12), 3062–3080 (2004)
8. T.S. Rappaport, R.W. Heath Jr., R.C. Daniels, J.N. Murdock, *Millimeter Wave Wireless Communications* (Prentice Hall, Upper Saddle River, 2014)
9. U. Kumar, V. Gupta, J.N. Laneman, Sufficient conditions for stabilizability over Gaussian relay and cascade channels, in *Proceedings of the IEEE Conference Decision and Control*, Atlanta, GA (2010), pp. 4765–4770
10. T.A. Johansen, A. Zolich, T. Hansen, A.J. Sørensen, Unmanned aerial vehicle as communication relay for autonomous underwater vehicle - field tests, in *Proceedings of the IEEE Globecom Workshop - Wireless Networking and Control for Unmanned Autonomous Vehicles*, Austin, TX (2014), pp. 1496–1474
11. S. Katti, H. Rahul, W. Hu, D. Katabi, M. Médard, J. Crowcroft, XORs in the air: practical wireless network coding. IEEE/ACM Trans. Netw. **16**(3), 497–510 (2008)
12. D. Hui, D.L. Neuhoff, Asymptotic analysis of optimal fixed-rate uniform scalar quantization. IEEE Trans. Inf. Theory **47**(3), 957–977 (2001)
13. R.M. Gray, D.L. Neuhoff, Quantization. IEEE Trans. Inf. Theory **44**(6), 2325–2383 (1998)
14. G. Caire, G. Taricco, E. Biglieri, Optimum power control over fading channels. IEEE Trans. Inf. Theory **45**(5), 1468–1489 (1999)
15. J.G. Proakis, *Digital Communications*, 4th edn. (McGraw-Hill, New York, 2001)
16. K. Gatsis, A. Ribeiro, G.J. Pappas, Optimal power management in wireless control systems. IEEE Trans. Autom. Control **59**(6), 1495–1510 (2014)
17. HART Communication Foundation, Control with WirelessHART (2009)
18. D.E. Quevedo, A. Ahlén, A.S. Leong, S. Dey, On Kalman filtering over fading wireless channels with controlled transmission powers. Automatica **48**(7), 1306–1316 (2012)
19. W.J. Dally, R.C. Harting, *Digital Design: A Systems Approach* (Cambridge University Press, Cambridge, 2012)
20. Z. Tang, H. Wang, Q. Hu, L. Hai, How network coding benefits converge-cast in wireless sensor network, in *Proceedings of the IEEE VTC Fall*, Quebec City, Canada (2012), pp. 1–5
21. G. Rajalingham, Q.-D. Ho, T. Le-Ngoc, Random linear network coding for converge-cast smart grid wireless networks, in *Proceedings of the QBSC*, Kingston, ON (2014), pp. 208–212
22. L. Schenato, Optimal estimation in networked control systems subject to random delay and packet drop. IEEE Trans. Autom. Control **53**(5), 1311–1317 (2008)
23. V. Gupta, A.F. Dana, J.P. Hespanha, R.M. Murray, B. Hassibi, Data transmission over networks for estimation and control. IEEE Trans. Autom. Control **54**(8), 1807–1819 (2009)
24. L. Shi, Kalman filtering over graphs: theory and applications. IEEE Trans. Autom. Control **54**(9), 2230–2234 (2009)
25. A. Chiuso, L. Schenato, Information fusion strategies and performance bounds in packet-drop networks. Automatica **47**, 1304–1316 (2011)

26. Y. Mo, E. Garone, A. Casavola, B. Sinopoli, Stochastic sensor scheduling for energy constrained estimation in multi-hop wireless sensor networks. IEEE Trans. Autom. Control **56**(10), 2489–2495 (2011)
27. L. Shi, A. Capponi, K.H. Johansson, R.M. Murray, Resource optimisation in a wireless sensor network with guaranteed estimator performance. IET Control Theory Appl. **4**(5), 710–723 (2010)
28. D.E. Quevedo, A. Ahlén, K.H. Johansson, State estimation over sensor networks with correlated wireless fading channels. IEEE Trans. Autom. Control **58**(3), 581–593 (2013)
29. E. Baskaran, J. Llorca, S.D. Milner, C.C. Davis, Topology reconfiguration with successive approximations, in *Proceedings of the MILCOM*, Orlando, FL (2007)
30. P. Agrawal, A. Ahlén, T. Olofsson, M. Gidlund, Long term channel characterization for energy efficient transmission in industrial environments. IEEE Trans. Commun. **62**(8), 3004–3014 (2014)
31. S.M. Ross, *Stochastic Processes*, 2nd edn. (Wiley, New York, 1996)
32. J.N. Al-Karaki, A.E. Kamal, Routing techniques in wireless sensor networks: a survey. IEEE Wirel. Commun. **11**(6), 6–28 (2004)
33. V. Pham, E. Larsen, K. Øvsthus, P. Engelstad, Ø. Kure, Rerouting time and queueing in proactive ad hoc networks, in *Proceedings of the IPCCC*, New Orleans, LA (2007), pp. 160–169
34. HART Communication Foundation, System redundancy with WirelessHART (2009)
35. B. Sinopoli, L. Schenato, M. Franceschetti, K. Poolla, M.I. Jordan, S.S. Sastry, Kalman filtering with intermittent observations. IEEE Trans. Autom. Control **49**(9), 1453–1464 (2004)
36. S. Dey, A.S. Leong, J.S. Evans, Kalman filtering with faded measurements. Automatica **45**(10), 2223–2233 (2009)
37. A.S. Leong, D.E. Quevedo, On the use of a relay for Kalman filtering over packet dropping links, in *Proceedings of the American Control Conference*, Washington, DC (2013), pp. 3320–3325
38. E. Rohr, D. Marelli, M. Fu, Statistical properties of the error covariance in a Kalman filter with random measurement losses, in *Proceedings of the IEEE Conference Decision and Control*, Atlanta, GA (2010), pp. 5881–5886
39. A.S. Leong, D.E. Quevedo, Kalman filtering with relays over wireless fading channels. IEEE Trans. Autom. Control **61**(6), 1643–1648 (2016)
40. A.S. Leong, D.E. Quevedo, A. Ahlén, K.H. Johansson, On network topology reconfiguration for remote state estimation. IEEE Trans. Autom. Control **61**(12), 3842–3856 (2016)
41. S.P. Meyn, Ergodic theorems for discrete time stochastic systems using a stochastic Lyapunov function. SIAM J. Control Optim. **27**(6), 1409–1439 (1989)

Chapter 6
Concluding Remarks

In this book, we have investigated the use of a number of different techniques from wireless communications, within the context of remote state estimation of dynamical systems. Techniques such as power control, energy harvesting, network coding, relays and rerouting have been introduced into remote state estimation problems. It has been shown that adapting appropriately to the time-varying wireless channel conditions, while taking into account the estimation quality, is usually beneficial and enhances system performance. Furthermore, in many of the problems studied in this book, optimal policies have been found to exhibit threshold-type structure, which simplifies their implementation.

As stated in the Introduction, one of the goals of this book is to bring closer together the wireless communications and control literature, by introducing wireless communications techniques and ideas into the study and design of networked control systems. We hope that by reading this book, the reader will have been inspired to investigate further into these approaches for use in networked estimation and control applications. We believe that many other wireless communication ideas and techniques, that have not treated in this book or elsewhere in the control literature, can also be successfully utilized by control researchers and practitioners.

© The Author(s) 2018   125
A.S. Leong et al., *Optimal Control of Energy Resources for State Estimation Over Wireless Channels*, SpringerBriefs in Control, Automation and Robotics, DOI 10.1007/978-3-319-65614-4_6